INDIVIDUAL BEHAVIOR AND COMMUNITY DYNAMICS

Population and Community Biology Series

Population and community biology is central to the science of ecology. This series of books explores many facets of population biology and the processes that determine the structure and dynamics of communities. Although individual authors have freedom to develop their subjects in their own way, these books are scientifically rigorous and generally adopt a quantitative approach.

Already published

7. **Stage-Structured Populations: Sampling, Analysis and Simulation**
   B.F.J. Manly (1990) 200pp. Hb.
8. **Habitat Structure: The Physical Arrangement of Objects in Space**
   S.S. Bell, E.D. McCoy and H.R. Mushinsky (1991, reprint 1994) 452pp. Hb.
9. **Dynamics of Nutrient Cycling and Food Webs**
   D.L. DeAngelis (1992) 285pp. Pb.
10. **Analytical Population Dynamics**
    T. Royama (1992 Hb, reprinted in Pb 1996) 387pp.
11. **Plant Succession: Theory and Prediction**
    D.C. Glenn-Lewin, R.K. Peet and T.T. Veblen (1992) 361pp. Hb.
12. **Risk Assessment in Conservation Biology**
    M.A. Burgman, S. Ferson and R. Akcakaya (1993) 324pp. Hb.
13. **Rarity**
    K. Gaston (1994) x+205pp. Hb/Pb.
14. **Fire and Plants**
    W.J. Bond and B.W. van Wilgen (1996) ca. xviii+263pp. Hb.
15. **Biological Invasions**
    M. Williams (1996) xii+244pp. Pb.
16. **Regulation and Stabilization: Paradigms in Population Ecology**
    P.J. den Boer and J. Reddingius (1996) xiii+397pp. Hb.
17. **Biology of Rarity**
    W. Kunin and K. Gaston (eds) (1997) Hb.
18. **Structured-Population Models in Marine, Terrestrial, and Freshwater Systems**
    S. Tuljapurkar and H. Caswell (eds) (1997) xii+643pp. Pb.
19. **Resource Competition**
    J.P. Grover (1997)
20. **Individual Behavior and Community Dynamics**
    J.M. Fryxell and P. Lundberg (1997)
21. **Spatiotemporal Models of Population and Community Dynamics**
    T. Czárán
22. **Dynamics of Shallow Lake Communities**
    M. Scheffer
23. **Chaos in Real Data: Analysis of non-linear dynamics from short ecological time series**
    J. Perry, D.R. Morse, R.H. Smith and I.P. Woiwood (eds)

INDIVIDUAL BEHAVIOR AND COMMUNITY DYNAMICS

John M. Fryxell
Department of Zoology, University of Guelph, Guelph, Ontario, Canada

Per Lundberg
Department of Theoretical Ecology, Ecology Building, Lund University, Lund, Sweden

CHAPMAN & HALL
London • Weinheim • New York • Tokyo • Melbourne • Madras

Printed in the United States of America

For more information, contact:

Chapman & Hall
115 Fifth Avenue
New York, NY 10003

Chapman & Hall
2-6 Boundary Row
London SE1 8HN
England

Thomas Nelson Australia
102 Dodds Street
South Melbourne, 3205
Victoria, Australia

Chapman & Hall GmbH
Postfach 100 263
D-69442 Weinheim
Germany

International Thomson Editores
Campos Eliseos 385, Piso 7
Col. Polanco
11560 Mexico D.F.
Mexico

International Thomson Publishing-Japan
Hirakawacho-cho Kyowa Building, 3F
1-2-1 Hirakawacho-cho
Chiyoda-ku, 102 Tokyo
Japan

International Thomson Publishing Asia
221 Henderson Road #05-10
Henderson Building
Singapore 0315

1 2 3 4 5 6 7 8 9 10 XXX 01 00 99 98

Library of Congress Cataloging-in-Publication Data

Fryxell, John M.
Individual behavior and community dynamics / John M. Fryxell and Per Lundberg.
p. cm. – (Population and community biology series ; 20)
Includes bibliographical references and index.
ISBN 0-412-99411-9 (alk. paper)
1. Animal behavior–Evolution. I. Animal ecology 3. Animal populations. 4. Predation (Biology) I. Lundberg, Per. II. Title. III. Series.
QL751.F878 1997
591.5–dc21
97-5547
CIP

British Library Cataloguing in Publication Data available

To order this or any other Chapman & Hall book, please contact **International Thomson Publishing, 7625 Empire Drive, Florence, KY 41042.** Phone: (606) 525-6600 or 1-800-842-3636.Fax: (606) 525-7778, e-mail: order@chaphall.com.

For a complete listing of Chapman & Hall's titles, send your requests to
Chapman & Hall, Dept. BC, 115 Fifth Avenue, New York, NY 10003.

Contents

Foreword

In Puccini's opera *Turandot*, the princes who court the daughter of heaven do so at their peril. Suitors of the Chinese princess must correctly answer three riddles. Answer one question incorrectly and the gentleman loses his life to the headman's sword. The severed heads of various failures decorate the stage as grim reminders of failed answers. Undeterred, Calaf, son of the dethroned Tartar king Timur, (ignoring the affections of the slave girl Liu, who loves him) strikes the gong and announces his great desire to solve the riddles and marry Turandot. And, being in the correct frame of mind, Calaf succeeds. In one last frenzy of drama, the pretty slave girl kills herself to save Calaf. At which point, Calaf and Turandot wed and presumably live happily ever after.

Evolutionary ecologists are suitors of knowledge regarding the principles and concepts that determine and predict the distribution and abundance of life. In a slight reversal of Puccini, we suitors, via our experiments and observations, ask questions of the ecological Turandot. Nature provides answers in the form of data. We ask questions at our peril. Misguided, squandered, or frivolous questions forfeit to the headman of ignorance. Ask the wrong question and we get an ambiguous answer at best or misunderstanding at worst. Just like the princes, we do not get to ask many questions in our lifetime of research. In our quest for understanding populations and communities of animals and plants, our questions must count.

John Fryxell and Per Lundberg have provided a wonderful service for evolutionary ecology. They have provided a comprehensive framework for thinking about populations and communities; a way of thinking that will make questions count. Ignore behaviors at one's peril is the message that John and Per bring to students, practitioners, and researchers of community ecology. Natural selection has imbued plants and animals with sensible behavioral responses to their environment, and these responses have profound consequences for the dynamics and structure of ecological communities. The eclipse of adaptive behaviors in most research on community ecology may have produced an abundance of severed heads decorating the stage. This message exists in many research papers and as a subtheme in book chapters and reviews. Scientists such as P. A. Abrams, Z. Abramsky, R. D. Holt, B. P. Kotler, R. H. MacArthur, D. W. Morris, L. Persson, M. L. Rosenzweig, E. E. Werner, and others are acknowledged for their significant contributions to this link between behaviors and communities. For the first time, however, this topic exists as the primary and comprehensive theme of a single treatise.

John and Per's treatise is crucial. When ingested as individual research papers, the topic can seem disjointed and confusing. Conflicting messages emerge for the role of adaptive behaviors in promoting species persistence, dynamical stability, and species diversity. John and Per clarify these issues beautifully. Conflicting messages emerge because adaptive behaviors can have diverse consequences—sometimes stabilizing sometimes not, sometimes promoting diversity sometimes not. Rather than capitulating to an anything goes conclusion, John and Per carefully circumscribe the conditions under which behaviors have profound effects (and they do!) and the association of these effects with ecological circumstances. Rather than try to accommodate a Babel of different models and modeling approaches that characterize the literature on the topic, John and Per have taken considerable effort to develop a single class of models and analyses for evaluating the full range of topics. This effort is amply rewarded. The chapters, each of which focuses on a different and significant ecological circumstance, flow seamlessly. This continuity of approach means that the book does not just review but presents original theory and ideas. As a textbook on behaviors and communities, the book's pedagogy nicely serves advanced undergraduate or graduate students. As a handbook, practitioners of ecology will find a useful guide for how to think about what behaviors and effects should be incorporated into management and conservation thinking. As a complete treatise, researchers will find coverage of the topics' "known world" and inspiration for future research. As a result of John and Per's effort, the future stage of community ecology should include better questions and fewer severed heads.

The imperative of John and Per's message may be greatest in the arena of human ecology and human impacts on biodiversity and the environment. Like Turandot's failed suitors, we ecologists have often been abysmal at predicting and influencing human impacts on the globe and its other denizens. Our pleas regarding the ills of overpopulation, overfishing, overconsumption, and overexploitation while certainly important, often sound shrill and lack relevance. What makes many ecologists sound like a chorus of Cassandras on issues in human economics, politics, and ecology? We all too frequently ignore the behavioral responses of Homo sapiens to their ecological and environmental circumstances. Ecologists often belittle or vilify technology, and as a result, lose perspective on present and future technologies as part of the human response to its environment. Fortunately, John and Per's theme can be profitably and aptly applied to humans as a species in the community. And, this fresh perspective should improve the questions ecologists ask and, consequently, amplify their influence. The ecological Turandot is ours for the asking—read on!

Joel Brown
Chicago, February 1997

Preface

This book arose from a series of casual conversations between Per and me in the spring of 1991, during his sabbatical leave at the University of Guelph. Comparing our respective experiences with different wildlife species, we realized that adaptive behavioral strategies seemed to underlie many of the feeding patterns we had observed. In trying to place those behavioral strategies in an ecological context, we were frustrated to find that the field had never received the kind of integrated treatment that would allow us to readily interpret our behavioral experiments. We decided that it would perhaps be useful to attempt such a unified treatment, linking a substantial body of previous theoretical and empirical studies.

It was another 3 years, however, before the opportunity presented itself for preparation of the book. Had we known how many subsequent complications there would be, I'm not entirely sure that we would have persisted, but persist we did. In this regard, Kjell Danell of the Swedish University of Agricultural Research was particularly instrumental, both through his encouragement of our collaborative research and through hosting my sabbatical leave in Umeå. Once we were in the same place at the same time, we couldn't think of any more excuses for procrastination and this book finally began to take form.

This seems an entirely appropriate place to express our appreciation to various individuals and organizations that have directly or indirectly contributed to our work over the years. Our respective research programmes have been generously supported by the following granting agencies: Natural Sciences and Engineering Research Council of Canada, Ontario Ministry of Natural Resources, Swedish Natural Science Research Council, Swedish Research Council for Forestry and Agriculture, Swedish Institute, and Royal Swedish Academy for Forestry and Agriculture. A number of individuals labored through early drafts of various parts of the manuscript: Andrew Illius, Os Schmitz, Ashley Mullen, Michael Usher, Emilio Laca, Kevin McCann, Tom Nudds, Peter Yodzis, Carey Bergman, John Wilmshurst, Daniel Fortin, and several anonymous reviewers. We thank them all for their careful criticism. My wife, Sue Pennant, took time from her busy schedule to apply her superb editing skills at several crucial points. Joel Brown, Burt Kotler, Mike Rosenzweig, Don DeAngelis, Don Kramer, Lennart Persson, Peter Abrams, Lauri Oksanen, Esa Ranta, Brad Anholt, Marc Mangel, Peter Yodzis, Tom Nudds, Kevin McCann, Mikael Sandell, Juha Tuomi, and Kjell Wallin provided valuable comments, discussion, and advice at various stages. Most of all, I thank my family for their patience while I was otherwise engaged.

Finally, I would like to take this opportunity to acknowledge the role of three key individuals in shaping my point of view as an ecologist. In an undergraduate course, Buzz Holling demonstrated the underlying beauty of biological theory. As I began my research career, Tony Sinclair taught me how to test ideas in an unforgiving world. Finally, at a later stage of my graduate studies, Carl Walters showed me how to link the two.

John Fryxell
Guelph, January 1997

In any intellectual endeavour, many people influence one's achievements. Many friends and colleagues at the places where I have been working have had their share in this project. Kjell Danell is the friend and colleague whose impact cannot be overestimated. Kjell also encourages a healthy suspicion of mathematical quirks. Mårten Åström was my collaborator during early stages of my work in this field and keen commentator at various stages of this work. Tom Nudds hosted me during my stay as visiting scientist at the University of Guelph. Tom made it possible for us to initiate this work and is therefore partially responsible for anything useful that may come of it. I also want to thank my wife and children for letting me do this and my father for teaching me everything.

Per Lundberg
Lund, January 1997

1 Introduction

Eat or be eaten: For many organisms, trophic interactions dominate the day-to-day struggle for genetic perpetuation. Not surprisingly, ecologists have devoted enormous amounts of time and energy to understanding this most fundamental of issues. Much of this work has relied on a dynamical systems approach pioneered by Lotka (1925) and Volterra (1928). The Lotka–Volterra equations were framed in the simplest possible manner, depicting ecological interactions as if they were simply random collisions between predators and their prey or their competitors. Although undeniably simplistic, the Lotka–Volterra approach has been enormously influential in guiding both the theoretical understanding of population dynamics and, to a lesser extent, interpretation of real population data (Kingsland 1985). Real organisms, however, have more complex patterns of resource consumption, movement, and intraspecific aggression—behavioral complications that are all too rarely incorporated into population theory.

Behavioral ecology focuses in great detail on the very aspects of individual behavior that population ecology ignores. Optimality theory (Stephens and Krebs 1986; Rosenzweig 1991) asserts that foraging behavior and habitat utilization are shaped by natural selection to maximize the fitness of individuals. An impressive body of empirical work has shown that simple optimality models of animal behavior are often useful in predicting the complex decisions made by real organisms (Stephens and Krebs 1986; Mangel and Clark 1988; Hughes 1990). In other words, evolutionary theory can be useful in mapping the relationship between individual behavioral or physiological performance and fitness. The next obvious mapping—between individual fitness-maximizing behavior and population dynamics—has received a good deal less attention. Our mission is to help rectify this imbalance by providing a broad introduction to adaptive behavioral strategies and their potential effect on population and community dynamics.

1.1 OBJECTIVES

In considering the implications of adaptive behavioral strategies, we explore a fundamental issue: whether adaptive behavioral decisions at the individual organism level tend to stabilize or destabilize trophic interactions. Classic ecological theory predicts that most interactions between predators and prey should be highly unstable or cyclical (Rosenzweig 1971; May 1973), regardless of whether the component species are plants, herbivores, carnivores, hosts, or parasitoids.

This prediction is inconsistent with our current information. Simple cycles do not seem to predominate in ecological communities (Witteman et al. 1990; Ellner and Turchin 1995), although there are certainly well-documented examples of population cycles (Turchin and Taylor 1992; Grenfell et al. 1992; Hanski et al. 1993; Sinclair et al. 1993; McLaren and Peterson 1994). This finding suggests that there are general features of predator–prey interactions that tend to stabilize interactions. It could be that ecological parameters promoting cyclic fluctuations are uncommon or it could be that unstable communities cycle so quickly to extinction that we rarely get a glimpse (Gilpin 1975; Yodzis and Innes 1992). On the other hand, it could be that natural selection tends to favor stabilizing mechanisms that counteract the intrinsic tendency of most trophic models for instability. If that were true, then behavioral interactions would seem a logical place to start, given that behavior is central to the question of who eats how much of whom.

In classic ecological theory, rates of consumption are commonly assumed to be a simple function of resource abundance, a relationship known as the functional response. Rates of consumption dictate the rate of population increase by predators as another simple function of resource abundance, a relationship known as the numerical response. Behavioral theory treats functional and numerical responses as being complex functions of such variables as the profitability of alternative resources, the spatial distribution of resources, or the social status and abundance of competitors.

We incorporate adaptive behavior into theoretical models by adjusting either the functional or numerical response, regardless of whether the decisions relate to diet breadth, patch use, social structure, or defensive behavior by prey. This provides a unified model structure; only the form of the modules specifying the functional and numerical responses vary across models. This modular approach should help readers with less modeling experience to concentrate on behavioral modifications rather than mastering a whole new set of equations with each change in topic. As a consequence of this modular design, we do not heavily emphasize previously published models that have fundamentally different structure, although we try to relate our findings to the literature whenever pertinent. This in no sense implies a negative value judgment about the quality of previous work; rather it emphasizes our belief that a common modeling structure might improve understanding for newcomers to the field.

Our choice of behavioral topics has been guided first and foremost by their immediate applicability to processes at the population level. We focus accordingly on behavioral aspects of foraging, space use, and aggressive interactions, ignoring a vast range of other fascinating behaviors that arise in other social contexts. We have striven to focus on behaviors that should apply across a wide range of taxa. Finally, we have chosen forms of behavior that have also received substantial theoretical and empirical interest in the behavioral ecology literature in hope that the kinds of behavior that have proved most interesting in laboratory settings might also prove most applicable to free-living populations.

In this book, we have chosen to be flexible in terms of the biological communities we have in mind. Some phenomena we discuss, such as adaptive partial predation or chemical defenses, are more pertinent to herbivores feeding on plants, whereas other adaptations relate more to classic carnivore–herbivore interactions. Although many of the concepts we consider should apply directly to host–parasite systems, particularly insect systems with parasitoid predators, we do not consider these specifically. The reason for excluding host–parasite models is simply that their inclusion would require a substantial increase in the number of models, without a substantial increase in conceptual understanding. Any of the models we present could be framed in classic Nicholson–Bailey form to mimic insect parasitoid–host systems.

We do not attempt here a comprehensive treatment of any single topic. For convenience, we often simply refer to generalized predator–prey or competitive interactions, rather than specifying which trophic levels are actually involved. Indeed, most of these topics are sufficiently rich to occupy entire research careers in themselves, particularly if one wished to actually test the mechanisms. Rather, we hope to provide a baseline for evaluating the range of population and evolutionary responses to common environmental constraints. Differences in methodology and jargon make it difficult for new researchers to bridge the subdisciplines of evolutionary and population ecology. At the very least, we hope to supply a useful guide for the uninitiated.

We have tried to illustrate our arguments with a liberal sprinkling of empirical examples. This book is scarcely the place for a detailed review, but we at least try to provide a reasonable introduction to the experimental literature. No doubt our choice of references are colored by our own background in terrestrial systems, so we apologize in advance for failing to recognize someone's favorite study. Behavioral ecologists often show impressive success in predicting decisions made by individuals, at least under controlled experimental conditions. The track record for population ecology pales by comparison, at least when it comes to empirical tests of predator–prey theory. None of our examples therefore provide a rigorous test of behavioral theory. Indeed, the current generation of models are so general that they no doubt have rather restricted predictive power with respect to any single system. It is nonetheless a worthwhile exercise to identify areas of consistency with the real world, that they/we may encourage a deeper, more sustained effort at model construction and falsification for specific systems (Rosenzweig 1991).

1.2 TOPICS TO BE COVERED

In the remainder of the first chapter we set the stage for the rest of the behavioral work by reviewing the inherent sources of instability of classic predator–prey models. The dynamical implications of behavior can be best understood by comparing them to classic models without behavior. Moreover, some of the principles of population regulation and trophic dynamics may not be familiar to students of behavioral ecology, nor the analytical methods commonly used in population ecology. We start with simple predator–prey interactions and then introduce exploitative competition at the upper trophic level.

In Chapter 2, we consider the classic foraging problem of adaptive diet choice among a range of species that differ in their energetic or nutritional characteristics. Diet choice is probably the oldest, if not the simplest, foraging problem. Its treatment should be familiar to most students of behavioral ecology. Nonetheless, there has been surprisingly little work done on the dynamical implications. In fact, only a handful of papers have treated this problem over the last two decades, despite the fact that most consumers depend on more than one food resource and that food selection has been shown to be nonrandom in virtually every case tested.

We explore the implications of defensive maneuvers by potential prey individuals in Chapter 3. Within this context, we consider two general types of defenses: risk-sensitive adjustment of activity levels by prey individuals and physical or chemical deterrents to reduce their attractiveness to predators. In each case, we assume life history trade-offs such that competitive ability or demographic characteristics are compromised to some degree by defense. As in all other chapters, we imbed such defensive maneuvers by prey individuals into trophic models, to explore the dynamical implications at the population level.

We consider the spatial distribution of resources and its effect on movement patterns in Chapter 4. Patch use has received a great deal of recent attention in the theoretical literature because of its application to habitat selection and metapopulation dynamics, areas of current emphasis in population ecology. Habitat selection theory rarely treats dynamics per se, and metapopulation theory often lacks an explicit evolutionary frame of reference for behaviors. For example, most metapopulation models assume constant probabilities of emigration (Kareiva 1990; Gilpin and Hanski 1991), whereas evolutionary models emphasize that emigration probability should be in some sense frequency and density dependent.

In Chapter 5, we consider the importance of adaptive responses to size variation among resources. Most consumers face a continuous range of ages, stages, or sizes of resources. If these size classes differ in energetic profitability and reproductive capacity, both of which seem likely, then there is potential for interesting dynamical effects. An adaptive predator could facultatively choose *which* prey to attack or choose *how much* of each item to consume once attacked. Both issues have interesting dynamical implications. The latter question should be particularly appropriate to plant–herbivore systems, because plants are rarely consumed entirely by their attackers. These systems also have the odd feature that biomass consumption by the herbivore and mortality in the plant population do not have the one-to-one mapping as in most other consumer–resource systems. The dynamical implications of this unique relationship are essentially unknown.

We explore the adaptive basis and dynamical implications of social structure within predator populations in Chapter 6. We first address direct aggressive interference among predators and its effects on rates of resource use. We then consider the effect of territorial spacing among predators and its effects on access to resources. We also consider the ecology of central place foraging, whether that central place might be a nest, den, or simply the center of a territorial home range. In each case, we consider the population and community implications arising from social heterogeneity among predators.

Chapter 7 offers some final comments on the value of an integrated view of behavioral ecology and population ecology. We consider causes of the current schism between researchers working at the scale of individual behavior and those interested in higher levels of organization. We also highlight topics that might be particularly fruitful for future work.

1.3 PREDATOR–PREY DYNAMICS

All population models assume that the rates of population change are governed by differences between per capita rates of recruitment and mortality. In Lotka–Volterra models, prey recruitment $G(N)$ is presumed to depend solely on prey density, which we symbolize using N. Prey mortality $X(N,P)$ is a direct function of consumption, which of course depends on both the density of prey N and the density of predators P. If the rate of energy gain influences the amount of secondary production of consumers, then it is reasonable to assume that predator recruitment $Y(N,P)$ is accordingly a function of both prey density and predator density. Mortality is assumed to depend solely on predator density $D(P)$. Dynamics of the simple community are depicted in terms of these general functions:

$$\frac{dN}{dt} = G(N) - X(N,P) \tag{1.1}$$

$$\frac{dP}{dt} = Y(N,P) - D(P) \tag{1.2}$$

The common linkage in all Lotka–Volterra models is therefore via the consumption and predator recruitment terms. We use the following set of equations (May 1973; Tanner 1975; Caughley 1976) as a null community model to which behavioral modifications will be added:

$$\frac{dN}{dt} = rN\left(1 - \frac{N}{K}\right) - \frac{aNP}{1 + ahN} \tag{1.3}$$

$$\frac{dP}{dt} = P\left(\frac{aceN}{1 + ahN} - d\right) \tag{1.4}$$

Note that the general Lotka–Volterra functions have now been given more explicit algebraic form. The resource growth function $G(N)$ is composed of two parameters: r = the maximum per capita rate of population growth and K = the resource carrying capacity. The consumption function $X(N,P)$ is composed of two parameters: a = the area searched per unit search time and h = the time required by predators to handle or process an individual prey item or a given biomass of resources. The predator growth function $Y(N,P)$ is composed of very similar terms as the consumption function, with the addition of the following extra parameters: c = a coefficient for converting energy consumption into predator offspring and e = the energy content of each individual prey item or unit biomass consumed. Finally, the function that predicts predator deaths $D(P)$ depends on one additional parameter: d = the per capita rate of predator mortality. We will always treat N and P as population densities (number of individuals per unit area), unless stated otherwise.

The dynamics of this particular formulation are quite well known (May 1973; Tanner 1975; Caughley 1976), which is one reason we use it as our point of departure. More importantly, however, this basic model also has two biological characteristics that should be part of most plausible models: density-dependent growth by resources and decelerating rates of consumption with increasing resource density. It is well worth taking some time to consider why these features are so important.

The prey growth function, which is called the logistic equation by population ecologists, stipulates that there is ultimately a limit on the resource abundance even when there are no predators present. The logistic equation ($rN[1 - N/K]$) depicts density-dependent limitation on the rate of population growth, using the simplest linear form one can imagine for a density-dependent relationship (Fig. 1.1*A*). The per capita rate of population growth is at a maximum when resource abundance is low and per capita growth declines proportionately with further increase in abundance. The net result, familiar to virtually any student of introductory population ecology, is a hump-shaped function of growth with respect to population abundance (Fig. 1.1*B*). The symmetry of the hump is a direct result of our simplistic assumption of linear density dependence. Other negative relationships between the per capita rate of growth and population density yield asymmetric hump-shaped functions. Literally dozens of studies show that per capita recruitment is negatively density dependent (Sinclair 1989), suggesting that the logistic equation provides a reasonable heuristic model of the rate of growth of at least some populations in the absence of predators.

The humplike form of the logistic recruitment function can be intuitively explained in the following way. The rate of change of any population depends on both the per capita rate of change ($r[1 - N/K]$) and the size of the population (N), regardless of whether the population one has in mind is the quantity of money in a bank account or a herd of wildebeest roaming the Serengeti Plains. When the per capita rate is small or the population is small, the rate of change will also be small. Rapid growth requires that both the per capita rate and the population density *both* be of adequate magnitude. According to the logistic model, at low population densities the per capita rate is high, but the capital on

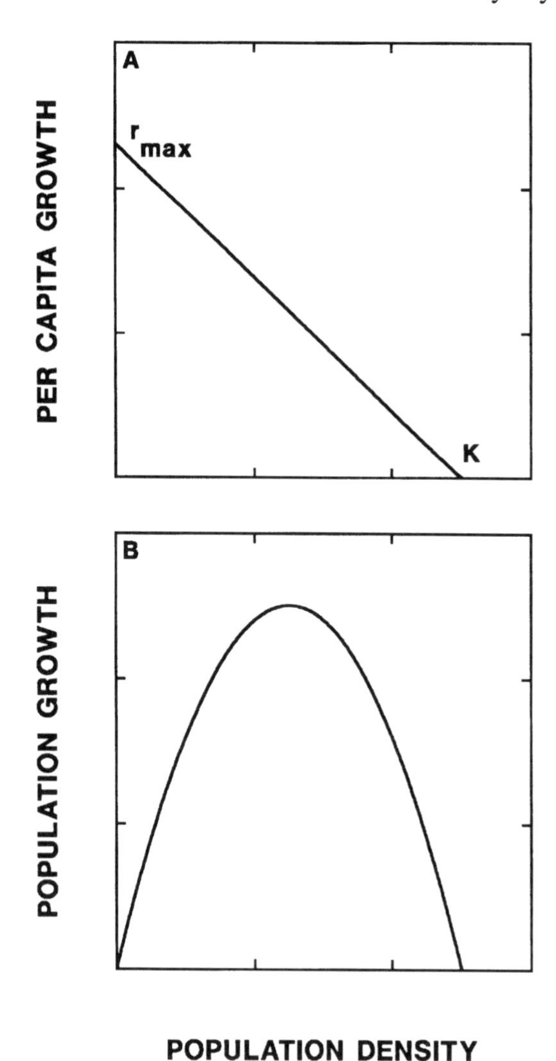

Figure 1.1. The rate of prey growth as a function of prey density (*A*, per capita growth; *B*, absolute growth) according to the logistic model. The rate of population growth is highest at intermediate population densities, because neither the per capita rate of change or population density are particularly small.

which this rate acts is small. Conversely, at high population densities the capital is large but the per capita rate is small. Hence, growth is fastest at intermediate population densities, specifically $K/2$.

The second term of the prey growth equation is the consumption rate by each individual predator ($aN/[1 + ahN]$), multiplied by predator density to calculate total losses to the prey population. The algebraic form of the consumption equations we have chosen to use is Holling's (1959) disc equation, so named because it was first tested using blindfolded human subjects "attacking" sandpaper discs scattered over a tabletop in a Canadian forestry lab. The humble origins of the disc equation belie its conceptual importance. It is a simple means of depicting changes in the rate of consumption in relation to changes in resource abundance (Fig. 1.2). The disc equation is often remarkably accurate in predicting consumption rates, no doubt because it mimics a general trade-off between search and destruction that applies to almost every predator, irrespective of whether that predator is a snake or a submarine.

According to the disc equation, consumption should rise initially with resource abundance at an approximately linear rate (aN). Most of the time, however, handling time inevitably affects the time left to search for new prey items, so there are diminishing returns as resources increase in abundance. One can readily see this by imagining the ridiculous circumstance in which a fox is literally knee-deep in rabbits. The maximum rate at which foxes could conceivably consume rabbits is obviously dictated by the time it takes to pop a new rabbit in its mouth, process it, and reach for another. At high prey densities, handling time therefore

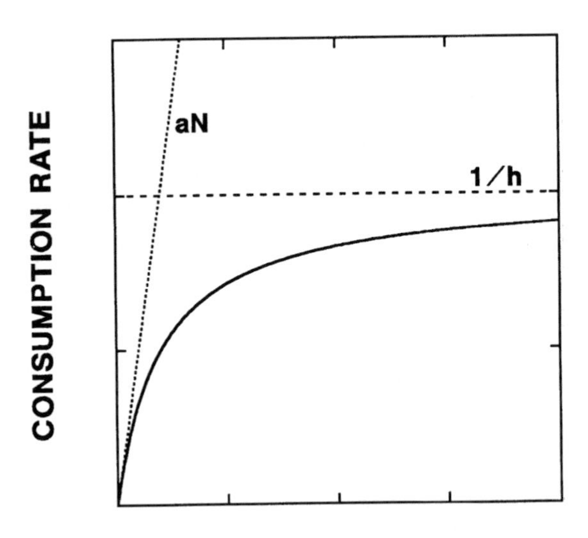

Figure 1.2. Rate of consumption as a function of prey density (termed the functional response), for a consumer faced with a single type of prey that requires a finite amount of time to process or handle. At low resource densities, consumption is constrained by the rate of search ($X \approx aN$), whereas consumption is constrained by the rate of handling resource items at high resource densities ($X \approx 1/h$).

constrains intake, such that consumption rates cannot exceed $1/h$ (Fig. 1.2). Hence, both the search rate (a) and handling time (h) influence consumption rates. There are literally hundreds of such functional response measurements, usually conducted with a single species of prey and a single species of predator and by far the majority exhibit the decelerating curve predicted by the disc equation.

The predator rate of increase is also presumed to be directly influenced by consumption rates, via conversion to new progeny. When consumption rates are low, due to resource scarcity, predators are presumed to decline exponentially at a per capita mortality rate of $-d$ (Fig. 1.3). When prey are more abundant, dP/dt rises with N, crossing a sustainability threshold (also known as the equilibrium prey density) at $N^* = d/(ace - adh)$. Under conditions of prey superabundance, the per capita rate of predator increase is ultimately limited to $(ce/h - d)$, due to the ultimate constraint on intake imposed by a finite handling time. Very few field studies have actually recorded per capita rates of change by predators in relation to resource density. There is at least one strong empirical study, however, showing that the per capita rate of population growth by red and grey kangaroos in relation to plant density was consistent with the model (Bayliss 1985).

By the same token, the rate of prey increase is contingent on consumption, which is determined by the functional response multiplied by predator density.

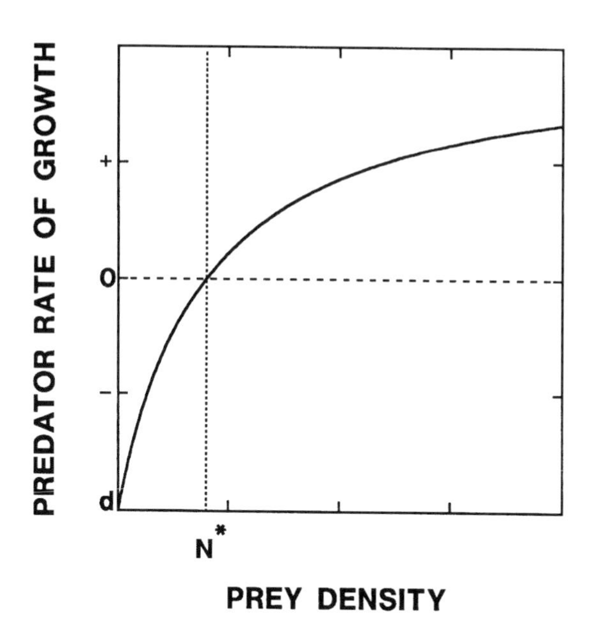

Figure 1.3. The per capita rate of change of predators (termed the numerical response) in relation to prey density, for a predator with a monotonically decelerating functional response.

Hence both prey and predator densities influence the per capita rate of increase by prey, but only prey density dictates the per capita rate of increase of predators. This is the other fundamental source of dynamical complexity arising from trophic interactions: The predator population is bound to track fluctuations in their resources.

The long-term dynamics of such systems depend on the magnitude of several key parameters (Caughley 1976), as is usually the case for more complex trophic interactions. In this case, however, the options are relatively limited in scope, comprising extinction of one component, stable equilibria of both components, or perpetual cycling of both components. There are several ways one can try to better understand how one set of equations can produce such different outcomes as recurrent cycles or constancy.

An intuitive way to understand the source of instability inherent in the Lotka–Volterra system is to consider the relative magnitude of density-dependent changes in prey recruitment due to intraspecific competition versus the inversely density-dependent risk of prey mortality due to predators. The logistic equation specifies that the per capita rate of prey recruitment declines with prey density according to $r(1 - N/K)$. Perturbation of a prey population to densities exceeding equilibrium would therefore cause a compensatory decrease in the per capita rate of recruitment, which acts as a negative feedback that would eventually reverse the perturbation. The per capita risk of mortality due to predation, $aP/(1 + ahN)$, declines with an increase in prey density, because consumption does not increase proportionately with resource density, but in a monotonically decelerating fashion. This mortality relationship implies that perturbation of the prey population to densities exceeding the equilibrium would cause an increase in the per capita rate of growth—a positive feedback leading to further divergence from equilibrium. All other things being equal, intraspecific competition within the prey population tends to restore equilibrium, whereas predation tends to destroy equilibrium.

The magnitude of the per capita rate of resource recruitment versus consumption term vary with the equilibrium density of prey, depending on parameter values of the system. The destabilizing effect of consumption often supersedes intraspecific competition at low prey densities, whereas the stabilizing effect of intraspecific competition often supersedes consumption at high prey densities. As a result, even simple predator–prey systems can be either stable or unstable, depending critically on the parameter values relevant to a given environment.

A physical metaphor can be used to demonstrate the importance of such a dynamic tension between stabilizing and destabilizing processes. Imagine a boy whirling around as he holds a string attached to a ball. Focus on the spatial position of the ball as the variable of interest. Two physical forces, centripetal force and gravity, interact to influence the position of the ball. When the ball is spun at a very slow rate, the stabilizing force of gravity supersedes the destabilizing centripetal force, with the result that the position of the ball changes little over time, tracing a tiny orbit around the boy. At high rates of spin, however, the destabilizing centripetal force supersedes the stabilizing force of gravity and the ball traces a wide orbit. This tension between stabilizing and destabilizing processes is a central tenet of population ecology, befitting the contemporary view of populations as parts of dynamic systems.

It is often useful to evaluate simple predator–prey models by examining the shape of their zero growth isoclines (Rosenzweig and MacArthur 1963) in the phase plane formed by their respective population densities. Zero growth isoclines are combinations of predator density and prey density at which one or the other population is unchanging (values of N and P at which $dN/dt = 0$ or $dP/dt = 0$). We do this by setting equations 1.3 and 1.4 to 0, and solving for N^* or P^*:

$$N^* = \frac{d}{a(ce - dh)} \tag{1.5}$$

$$P^* = \left(\frac{r}{aK}\right)(K - N)(1 + ahN) \tag{1.6}$$

Note that the equilibrium density of prey is a constant, whereas the equilibrium density of predators changes with prey population density.

When plotted in the predator–prey phase plan, the predator zero isocline is vertical, intersecting the horizontal axis at the prey density that is just sufficient to match consumer recruitment with predator mortality (N^* in Fig. 1.3). In contrast, the prey zero isocline is a hump-shaped function of prey population density. The ascending limb of the hump corresponds to conditions under which destabilizing effects of consumption supersede the stabilizing effects of intraspecific competition, whereas the opposite is true of the descending limb of the prey zero isocline (Rosenzweig and MacArthur 1963). The model community would be stable if parameters are such that the predator isocline intersects the prey isocline to the right of the hump, where the negative slope of the zero isocline implies negative feedback on prey population growth near the joint equilibrium (Fig. 1.4*A*). The model community would be unstable if the predator isocline intersects the prey isocline to the left of the hump, where the positive slope of the prey isocline implies a positive feedback on resource population growth near the joint equilibrium (Fig. 1.4*B*).

A mathematically rigorous approach to understanding these processes is to apply local stability analysis. There are several excellent explanations of local stability as it is applicable here (May 1973; Nisbet and Gurney 1982; Pimm 1982; Yodzis 1989), so we will skip most of the technical details and concentrate more on its intuitive application to predator–prey dynamics. Any readers who find this heavy going can readily skip to the next section on competition with little loss of intuitive understanding.

Local stability analysis explores the short-term response of the community when it is slightly displaced from equilibrium conditions. Not surprisingly, one

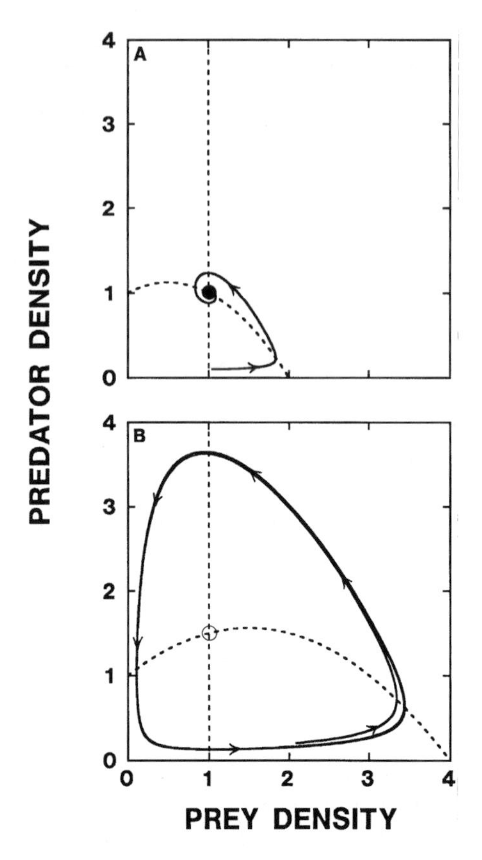

Figure 1.4. Stable (*A*) and unstable (*B*) population trajectories for predators and prey for the Lotka–Volterra model ($a = h = r = d = 1$, $e = 2$, $K = 2$ in Fig. 1.4*A* and $K = 4$ in Fig. 1.4*B*). Zero isoclines for predators (vertical line) and prey (hump-shaped curve) are depicted by dotted lines. Nontrivial equilibria are shown by symbols (filled = stable, open = unstable). Note that the system is stable in the case in which the prey zero isocline has a negative slope at the intersection with the predator zero isocline (*A*), whereas the system is unstable where the slope of the prey isocline is positive at the intersection with the predator isocline (*B*).

must therefore first find this equilibrium (or equilibria if there is more than one). These are combinations of prey and predator densities at which each population is unchanging (N^* and P^* at which both $dN/dt = O$ and $dP/dt = O$). For the models we will consider there is usually only one equilibrium, but more equilibria would be possible for more complex models.

The second step of local stability analysis is to calculate the effects of slight perturbations to first N and then P on the rates of population growth, dN/dt and dP/dt. Although the rate of growth of any given population obviously depends on the population densities of other species in the system, we can proceed in steps by first calculating the partial derivatives of each population growth equation in relation to change in each other species in the community. Consider the predator–prey model depicted in equations 1.3 and 1.4 as an example:

$$\alpha_{NN} = \frac{\partial(dN/dt)}{\partial N} = r - \frac{2rN^*}{K} - \frac{aP^*}{(1 + ahN^*)^2} \tag{1.7}$$

$$\alpha_{NP} = \frac{\partial(dN/dt)}{\partial P} = - \frac{aN^*}{(1 + ahN^*)} \tag{1.8}$$

$$\alpha_{PN} = \frac{\partial(dP/dt)}{\partial N} = \frac{aceP^*}{(1 + ahN^*)^2} \tag{1.9}$$

$$\alpha_{PP} = \frac{\partial(dP/dt)}{\partial P} = \frac{aceN^*}{1 + ahN^*} - d = 0 \tag{1.10}$$

The α-values are known as the community matrix coefficients. When equilibrium densities of predators and prey are inserted, these become numerical constants that simply describe the rate of growth (or decline) with an incremental increase in density of a given population. The community coefficient α_{NN} represents the effect of a small increase in prey density on the rate of prey population growth, α_{NP} represents the effect of a small increase in predator density on the rate of prey population growth, α_{PN} represents the effect of a small increase in prey density on the rate of predator population growth, and α_{PP} represents the effect of a small increase in predator density on the rate of predator population growth. Consider α_{NP}, for example. A negative value of α_{NP} implies negative feedback between predator density and prey population growth. By definition, we know that $dN/dt = 0$ at equilibrium, so a negative α_{NP} indicates that a prey population would tend to return to its former equilibrium should it ever be slightly perturbed from that equilibrium. In any kind of dynamic system, negative feedback is essential for local stability, or natural regulation of abundance as it is called by many ecologists (Sinclair 1989).

Negative feedback on any single population may not be sufficient to guarantee stability, if there are other ecological interactions in the community. Positive feedbacks can potentially destabilize the community if they are larger than the negative feedbacks. The only way we can know for sure is to combine community coefficients for each population and then approximate the dynamics of the entire system near equilibrium.

The rate of change of deviations between the population density and equilibrium density (symbolized by $\underline{N}^* = N(t) - N^*$ and $\underline{P} = P(t) - P^*$) can be estimated to a first approximation by the following first terms of the Taylor expansions:

$$\frac{d\underline{N}}{dt} = \alpha_{NN}\underline{N} + \alpha_{NP}\underline{P} \tag{1.11}$$

$$\frac{d\underline{P}}{dt} = \alpha_{PN}\underline{N} + \alpha_{PP}\underline{P} \tag{1.12}$$

These equations indicate that the complicated ecological interactions can be approximated *near equilibrium* by the additive effect of the community matrix coefficients. Each of the terms in these linear equations are composed of 2 elements: a numerical coefficient (α_{ij}) multiplied by the deviation from equilibrium ($\underline{N}$ or $\underline{P}$). This linear approximation is very useful because linear differential equations such as equations 1.11 and 1.12 can be solved directly. The solution will generally be of the form

$$\underline{N}(t) = \kappa_1 e^{\lambda_1 t} + \kappa_2 e^{\lambda_2 t} \tag{1.13}$$

$$\underline{P}(t) = \kappa_3 e^{\lambda_1 t} + \kappa_4 e^{\lambda_2 t} \tag{1.14}$$

where κ_1 through κ_4 are constants that correspond to the magnitude of the initial perturbations from equilibrium and the eigenvalues λ_1 and λ_2 are the roots of the characteristic polynomial:

$$\lambda^2 - (\alpha_{NN} + \alpha_{PP})\lambda + (\alpha_{NN}\alpha_{PP} - \alpha_{NP}\alpha_{PN}) = 0 \tag{1.15}$$

A careful examination of the linearized equations 1.13 and 1.14 indicates that $\underline{N}(t)$ and $\underline{P}(t)$ will converge on 0 (that is $N(t) \rightarrow N^*$) and $P(t) \rightarrow P^*$) if λ_1 and λ_2 are negative, but will grow even larger if $\lambda_1 > 0$, $\lambda_2 > 0$, or both. In our specific example, λ_1 and $\lambda_2 < 0$ so long as $\alpha_{NN}\alpha_{PP} - \alpha_{NP}\alpha_{PN} > 0$ and $\alpha_{NN} + \alpha_{PP} < 0$.

We have already explained why increased prey density would usually increase the rate of energy gain by consumers. As a consequence, $\alpha_{PN} > 0$ in the simple system described by equation 1.3 and 1.4, provided that the system can sustain a positive predator population, i.e., $K > d/a(e - dh)$. By the same token, $\alpha_{NP} < 0$ in the simple predator–prey system. Multiplying each of these terms together, $\alpha_{NP}\alpha_{PN} < 0$ for all communities capable of sustaining predators. By definition, $\alpha_{pp} = 0$. Hence, local stability of the community depicted by equations 1.3 and 1.4 depends on the sign of α_{NN}. This can be either positive or negative, depending on the magnitude of $r(1 - 2N^*/K) - aP^*/(1 + ahN^*)^2$. In an environment supporting a low carrying capacity of prey, $2rN^*/K + aP^*/(1 + ahN^*)^2$ tends to be large relative to r and the system tends to be stable. In a highly productive environment, $2rN^*/K + aP^*/(1 + ahN^*)^2$ tends to be smaller than r and the system is correspondingly unstable.

Stability analysis yields useful insights into the effects of changing environmental circumstances, because the community matrix coefficients (α_{NN}, α_{NP}, α_{PN}, and α_{PP}) are themselves functions of community parameters (*a, h, c, e, d, r,* and *K* in equation 1.1). For example, Figure 1.5 shows the range of parameter combinations for energy content versus prey carrying capacity at which (1) predators cannot coexist with their prey, (2) prey coexist stably with their predators, and (3) predators and prey both cycle endlessly from high to low abundance and back again. The progression from one state to the next as one traces parameter combinations outward from the origin gives some intuitive biological basis to the mathematics. Systems in which prey are rare even in the absence of predators (small *K*) or of extremely poor quality (small *e*) are unable to sustain predators permanently. Conversely, systems in which resources are extremely common or of extremely high quality tend to encourage oscillations due to rather weak density dependence in the region of equilibria. Hence, predators continue to increase even as they deplete resources to low densities, because even low rates of consumption are sufficient to allow some predator recruitment. Prey continue to increase beyond equilibrium when predators are rare, because the intrinsic

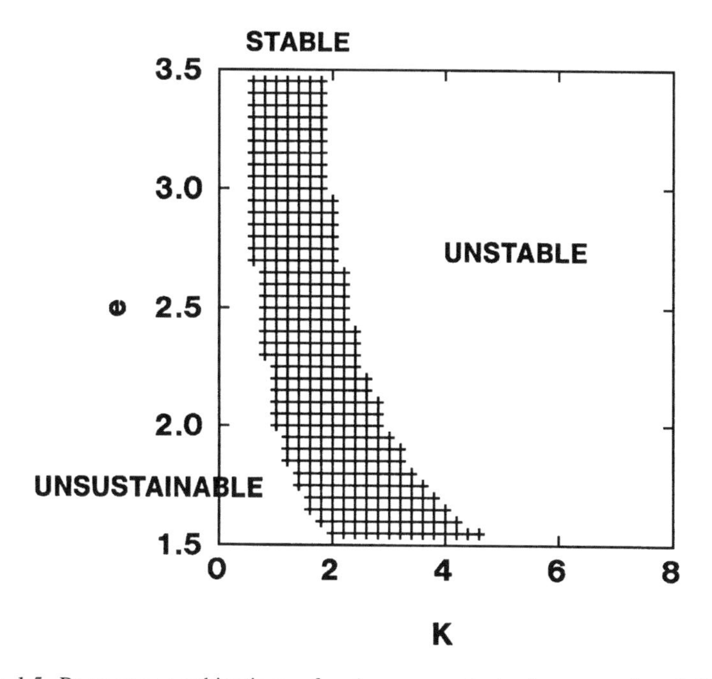

Figure 1.5. Parameter combinations of *e* (energy content of resources) and *K* (prey carrying capacity) leading to stable coexistence of predators and prey (+). Parameter combinations to the left of the stable range are insufficient to allow predators to persist, whereas parameter combinations to the right of stable range are dynamically unstable, producing recurrent oscillations in densities of predators and prey. Other parameter values are as follows: $a = h = r = d = 1$.

density dependence of prey (arising from the logistic recruitment equation) is weaker than the inverse density dependence imposed by consumers (arising from the decelerating functional response). The latter condition has been termed the "Paradox of Enrichment," in which improved supplies of the critical resource for consumers may lead to unstable population dynamics and even extinction due to demographic stochasticity at troughs in the cycle (Rosenzweig 1971; Gilpin 1972; May 1972).

The most important reason, therefore, for evaluating the stability of simple community models is that it gives an appropriate yardstick for assessing the effects of behavioral modifications or other changes to model components. According to this view, the Lotka–Volterra equations are the mathematical analogue of the white lab rat. Tinkering with the basic model informs us whether a particular biological consideration, such as individual behavior, *might* have interesting implications for more realistic models, judging from its effect on a commonly accepted (but less realistic) standard. We rely on this theme heavily throughout the book.

Evaluation of local stability becomes considerably more problematic, however, with each stepwise increase in dimensionality (i.e., the number of separate subpopulations of interest). The behavioral models we discuss usually involve > 3 sets of populations or subpopulations, at which point analytical methods of predicting local stability border on masochism. As a brute force alternative, population ecologists are usually forced to rely on computer simulations to estimate population variability over time for a given range of parameter values (Fig. 1.6). This means of evaluating stability is still valid (compare Figs. 1.5 versus 1.6), but it is certainly less elegant than a detailed analysis of local stability. Unfortunately, one has little choice when confronted by ecological models of even intermediate complexity.

We generally assume a constant environment in our models, if only to simplify the unmanageable. This assumption flies in the face of the practical experience of every empiricist: There is no such thing as a purely deterministic population interaction, even in the lab. Nonetheless, deterministic models can still yield useful insights into the behavior of more realistic, noisy systems. It is helpful, nevertheless, to remember that the patterns that we might expect to see in nature are often highly fragmented and obscured by stochastic sources of variability beyond our comprehension.

1.4 COMPETITION

From the beginning, ecologists have recognized that interactions within food webs extend horizontally as well as vertically, i.e., competition among consumers of a common resource could be as important as a determinant of community dynamics as simple links between predators and their prey. The simplest theoretical approach to this problem is to assume that nonliving, yet constantly renewing resources instantly adjust to changes in the population densities of competing

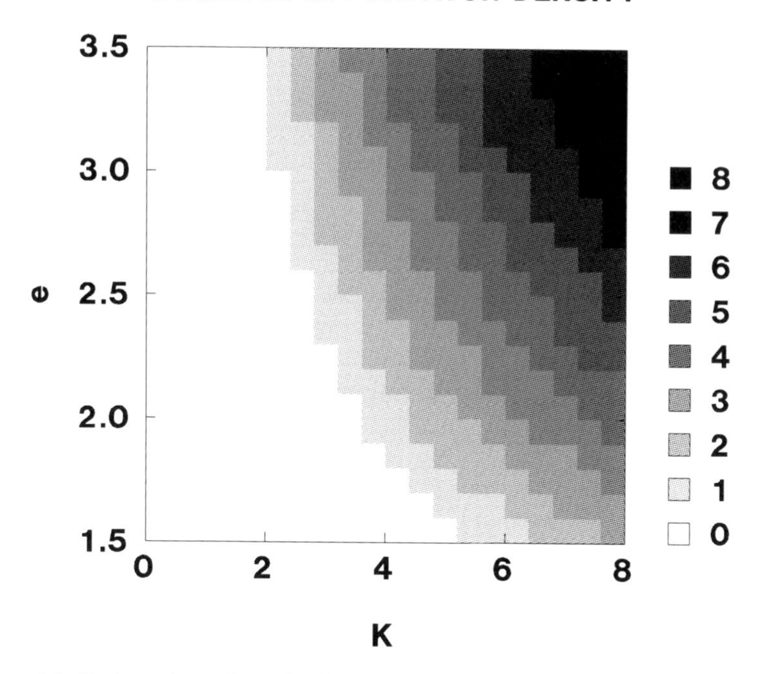

Figure 1.6. Variance in predator density recorded over the last 500 time steps of a 1000 step simulation, for the same parameter combinations shown in Figure 1.5.

consumers. If prey items have an infinitesimally small handling time, leading to a linear functional response, then competing consumers might subdivide food as quickly as it is provided. Under these rather special circumstances, community interaction can be distilled to the following pairwise interaction between competitors (Yodzis 1989, pp. 120–121):

$$\frac{dN_1}{dt} = r_1 N_1 \left(\frac{K_1 - N_1 - b_{12} N_2}{K_1} \right) \tag{1.16}$$

$$\frac{dN_2}{dt} = r_2 N_2 \left(\frac{K_2 - N_2 - b_{21} N_1}{K_2} \right) \tag{1.17}$$

where r_i = maximum per capita rate of increase of competitor species i and $b_{i,j}$ = the depressant effect of a single unit of population density of competitor species j on the per capita rate of increase of species i, which is often termed the competition coefficient in ecological jargon. This classic model of competition is referred to repeatedly in legions of introductory ecology textbooks.

These dynamic outcomes can once again be interpreted in relation to their zero isoclines, which are calculated as combinations of N_1 and N_2 at which $dN_1/dt = 0$ or at which $dN_2/dt = 0$:

$$N_2 = \frac{K_1}{b_{12}} - \frac{N_1}{b_{12}} \tag{1.18}$$

$$N_2 = K_2 - b_{21}N_1 \tag{1.19}$$

The zero isoclines are straight lines plotted in the phase plane of N_1 versus N_2. The zero isocline for species 2 has an intercept of K_2 on the N_2-axis and a slope of $-b_{21}$, whereas species 1 has an intercept of K_1/b_{12} on the N_2-axis and a slope of $-1/b_{12}$. When species 1 is a strong competitor whereas species 2 is a weak competitor, then the trajectory shoots off to the equilibrium K_1 (Fig. 1.7*A*). This means, of course, that the second species is driven extinct. On the other hand, parameter values implying that species 2 is a strong competitor whereas species 1 is a weak competitor produce a trajectory that leads to K_2 (Fig. 1.7*B*).

If both competitors exert a stronger density-dependent influence on the rate of increase of heterospecifics than conspecifics, then the outcome depends critically on initial conditions of the competition experiment (Fig. 1.7*C*). If both species are strong competitors and species 1 has a substantial initial numerical advantage, then species 1 excludes 2. Conversely, if species 2 has the initial advantage, then species 2 vanquishes species 1. Finally, if both species are less strong in competition with the other species than when competing with conspecifics, then the outcome is a stable equilibrium (Fig. 1.7*D*).

This model of pairwise competition has exactly the same limitations, unfortunately, as the logistic model of single population growth. For the same reasons that a predator–prey model is intrinsically more "realistic" than a single-species logistic model, mechanistic models of heterotrophic competition are badly needed before one can lay claim to understanding the structure and dynamics of even simple communities (Schoener 1986).

It is simple to expand the basic predator–prey model outlined earlier to mimic exploitative competition. One simply specifies a single resource of density N, preyed upon by > 1 species of consumer of population density P_i. For 2 competitors, the full mechanistic model might be depicted in the following way:

$$\frac{dN}{dt} = rN\left(1 - \frac{N}{K}\right) - \sum_{i=1}^{2} \frac{a_i N P_i}{1 + a_i h_i N} \tag{1.20}$$

$$\frac{dP_1}{dt} = P_1\left(\frac{a_1 c_1 e N}{1 + a_1 h_1 N} - d_1\right) \tag{1.21}$$

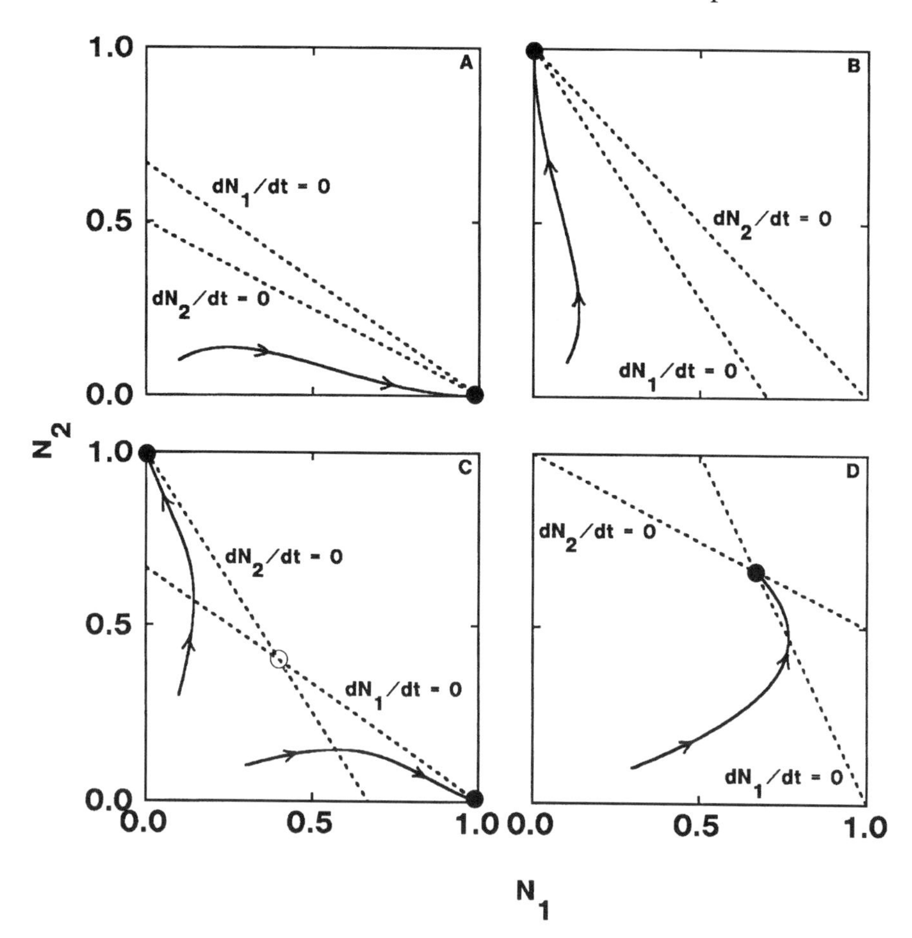

Figure 1.7. Four outcomes of Lotka-Volterra competition between two species: (*A*) species 1 excludes species 2, (*B*) species 2 excludes species 1, (*C*) either species wins, depending on initial conditions, and (*D*) both species coexist at equilibrium. The straight lines correspond to zero isoclines of each species; the curved lines correspond to trajectories over time of both competitors. Nontrivial equilibria are shown by symbols (filled = stable, open = unstable). The following parameter values were used for the simulations: (*A*) $K_1 = 1$, $K_2 = 0.5$, $b_{21} = 0.5$, $b_{12} = 1.5$; (*B*) $K_1 = 0.7$, $K_2 = 1$, $b_{21} = 1.0$, $b_{12} = 0.7$; (*C*) $K_1 = 1$, $K_2 = 1$, $b_{21} = 1.5$, $b_{12} = 1.5$; (*D*) $K_1 = 1$, $K_2 = 1$, $b_{21} = 0.5$, $b_{12} = 0.5$.

$$\frac{dP_2}{dt} = P_2\left(\frac{a_2c_2eN}{1 + a_2h_2N} - d_2\right) \tag{1.22}$$

The range of outcomes of such three-dimensional models differs somewhat to that of the pairwise competition model, ranging from competitive exclusion of one or the other competitor to perpetual coexistence of both competitors (Koch 1974; Hsu et al. 1978; Armstrong and McGehee 1980; Waltman 1983). An example illustrates these possibilities. Consider 2 competing species of predators, one of which has a high rate of search but a prolonged handling time and a second with a low rate of search but a short handling time. One could perhaps think of these organisms as either being adapted to maximize search efficiency or processing efficiency. As a consequence, efficient searchers would have higher per capita rates of growth than efficient handlers at low prey densities, whereas the converse would be true at high prey densities (Fig. 1.8).

At low prey carrying capacities, one or the other species inevitably excludes the other. The reason for this should be intuitively obvious—consumers with different numerical responses by definition have different sustainability thresholds. The predator species that can sustain itself at the lower resource density will usually squeeze its competitor out. If the numerical responses intersect at a resource density well below either of the equilibria, then the efficient handler wins the interaction (Fig. 1.9*A*), because it maintains a higher rate of energy gain than the efficient searcher at all relevant densities. Should the crossover occur above the equilibria, then the efficient searcher wins the interaction.

However, one other important outcome is possible: Both competitors can

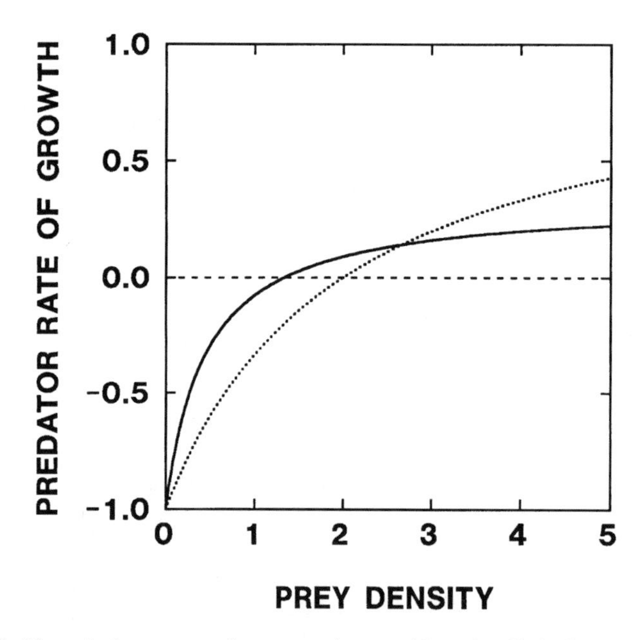

Figure 1.8. Numerical responses for two predators with trade-offs in foraging efficiency at low versus high prey densities. One species is more efficient than the other at foraging at low prey densities ($a_1 = 3$, $a_2 = 1$), whereas the converse is true at high prey densities ($h_1 = 0.75$, $h_2 = 0.5$). The dotted horizontal line depicts equilibrial conditions ($dP_i/P_i dt = 0$).

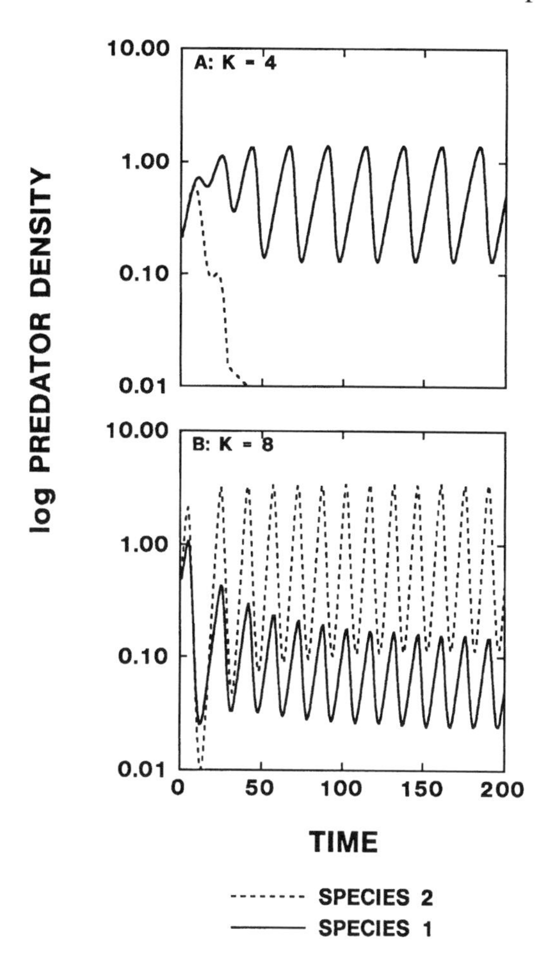

Figure 1.9. Changes in competitor densities over time in purely exploitative competition in depauperate environments (*A*: $K = 4$) versus enriched environments (*B*: $K = 8$). Functional response parameters as in Figure 1.8, all other parameters as follows: $r = c_1 = c_2 = e = d_1 = d_2 = 1$). Note that both competitors coexist only in the enriched system.

coexist, exhibiting population cycles that are slightly out of phase. For example, if the efficient handler tends to win at low K, then at higher values of K both competitors can coexist (Fig. 1.9*B*). Competitive exclusion can even be reversed for some parameter combinations (Hsu et al. 1978). Under this purely exploitative form of competition, dynamic instability is essential for competitive coexistence (Armstrong and McGehee 1980; Waltman 1983). Competitive coexistence under these exploitative conditions derives from the fact that resource density can vary enough over the full cycle to guarantee that both competitors have identical mean fitness. Hence, competitive coexistence can occur when there are trade-offs in foraging efficiency at low versus high resource densities, as in Fig. 1.8, *and* the community is unstable (Armstrong and McGehee 1980). In conclusion, horizontal (competitive) food web interactions can be approached in an analogous fashion to vertical (predator–prey) interactions.

Instead of assuming that competing predators represent different species, one might just as legitimately regard them as different behavioral genotypes in a single predator population. Without further specification of genetic structure, the predator genotypes are perhaps best regarded as asexual clones. In a landmark paper, Maynard Smith and Price (1973) used game theory to define the useful concept of an evolutionarily stable strategy (ESS). An ESS is a strategy such that if most of the members of a population adopt it, there is no mutant behavior yielding greater fitness. In order to qualify as an ESS, therefore, a behavioral strategy must exhibit two properties (Maynard Smith 1982). First, expected fitness of an ESS when it is rare in a population must be equal or higher to that of any other genotypes that might be present (sometimes termed the stability condition). Second, once fixed in the population, mean fitness of the ESS must exceed or be equal to that of any mutant genotype (sometimes termed the equilibrium condition). The exploitative competition model makes the valuable point that unstable trophic communities might support polymorphic consumers, with different segments of the population being most successful at different phases of the population cycle. Recent work on unstable grass-sheep communities on the Island of St. Kilda is consistent with this scenario. Sheep on St. Kilda are highly unstable, exhibiting cyclical fluctuations in density with a 3-year period (Clutton-Brock et al. 1991; Grenfell et al. 1992). Strong selection occurs during population crashes, yet wide variability in jaw morphology and feeding efficiency persists in the face of intense natural selection, presumably because of compensatory fitness advantage at other phases of the population cycle (Illius et al. 1995). We discuss the community ESS at various stages in this book to consider the implications of community dynamics on natural selection of genotypes with different foraging behavior.

1.5 SUMMARY

We have tried to explain why a consideration of behavior in relation to community ecology should be of paramount interest and to sketch out the organizational details of our own attempt at such a synthesis. There are two biological characteristics that we feel ought to be part of any general theory of community dynamics. One unavoidable fact is that no population should be capable of infinite growth, even in the absence of a higher consumer. Another biological necessity is that consumption rates must eventually have limits, implying that consumption cannot keep pace with incremental changes in prey population abundance. A natural consequence of these biological characteristics is that stability of predator and prey populations is an unlikely proposition, yet there is not overwhelming empirical evidence that population cycles are the norm rather than the exception. We suggest that natural selection for patterns of predator and prey behavior that optimize individual fitness might be the kind of general process that could explain this apparent paradox.

To understand the effect of behavior on community dynamics and, conversely, the effect of community dynamics on the evolution of behavior, one needs a benchmark against which system modifications can be assessed. Modified Lotka–Volterra models of predator–prey dynamics and mechanistic competition offer a simple theoretical template for addressing these questions. If consumption is monotonically decelerating and prey have logistic limits to growth, then the Lotka–Volterra equations exhibit capacity for either stable or unstable dynamics, contingent on parameter values relevant to a given system. Based on previous ecological studies, we consider in some detail the biological sources of instability inherent in these relationships, discussing local stability, qualitative interpretation of dynamics via isocline analysis, and the necessity of using numerical simulations in order to understand moderately detailed behavioral models. We also use previous work on mechanistic interspecific competition to show how community instability can favor maintenance of alternate predator genotypes in the same environment, setting the scene for considerations of the evolution of adaptive behavior in later chapters.

2 Diet Selection

Animals are faced with an astonishingly wide variety of things to eat, and each of these foods has an equally wide variety of nutritional constituents. One might suppose accordingly that natural selection would favor wise patterns of diet selection and that the principles guiding diet selection would be relatively straightforward. Several generations of biologists have addressed this very issue, most recently and successfully by borrowing from economics the concepts and procedures of optimality theory (Belovsky 1986; Schoener 1986; Stephens and Krebs 1986). Although it would be naive to claim a consensus has emerged (Pierce and Ollason 1987; Stearns and Schmid-Hempel 1987; Ward 1992, 1993; Nonacs 1993), optimality models of diet selection have received wide acceptance.

A central assumption of optimal foraging theory is that different food items potentially available to any predator (regardless of whether that predator is carnivorous, granivorous, or herbivorous) differ with respect to their intrinsic profitability to that predator. In standard ecological jargon, profitability is precisely defined as the nutrient or energy content of a single typical prey item divided by the time it takes for a predator to handle that item. Hence, profitability is synonymous with the maximum instantaneous rate of resource gain under perfect conditions, i.e., when a predator already has a prey item in its grasp and only faces the further time investment in consuming and processing that item before reaping its intrinsic reward. It is therefore not surprising that virtually all diet selection models involve comparisons among items with different relative profitability rankings (Stephens and Krebs 1986). Other factors, such as risk of predation or habitat structure, could constrain diet selection, but for simplicity sake we start with a simplistic scenario in which prey items are homogeneously distributed and a forager is only concerned with maximizing the rate of energy gain.

It is well worth emphasizing that profitability can differ between two items on the basis of differences in their energy content, handling time, or both. Energetic variability has received the greatest attention from ecologists, no doubt because it is rather more simple and rather less messy to incinerate tissue samples in a bomb calorimeter than to record all the gritty details of handling time, including progression through the digestive tract. Nonetheless, there is evidence supporting the importance of handling time in shaping diet selection (e.g., Belovsky 1986; Bjorndal 1991; Kaspari and Joern 1993; Scheel 1993; Doucet and Fryxell 1993). Handling time is most likely to be an issue for herbivores, simply because there is often pronounced variation in the time it takes for different plant tissues to be digested and cleared from the digestive tract.

Out of the wide variety of theoretical models that address optimal diet selection, we concentrate on variants of the earliest version, the contingency model (MacArthur and Pianka 1966; Schoener 1971; Pulliam 1974, 1975; Charnov 1976*b*; Belovsky 1978; Stenseth and Hansson 1979; Stenseth 1981; Owen-Smith and Novellie 1982; Engen and Stenseth 1984; Stephens and Krebs 1986; Verlinden and Wiley 1989). Our rationale is simple: The central elements of virtually all succeeding generations of diet choice models are embodied in the contingency model, the structure is simple enough to be recognizable when inserted into classic models of predator–prey and competitive interactions, and the structure of the behavioral model can be readily adapted to different biological circumstances. The latter point is of particular importance, because in keeping with our "white rat" view of theoretical models, we stress the ecological importance of biological differences between models, embodied by simple changes to model structure or parameter values. We compare systems in which prey are additive in their effects (no nutritional synergism) versus systems in which prey have complementary or antagonistic nutritional benefits to the consumer.

2.1 NUTRIENT-MAXIMIZING DIETS

The underlying premise of optimal foraging theory is that fitness is a direct consequence of the rate of nutrient or energy gain. Perhaps the simplest generic case one can imagine is that all prey available to the predator are randomly distributed in space and that the nutritional value of each prey item is independent of that of any other items in the predator's diet (sometimes termed nutritional additivity). Further assume that there are 2 prey species of interest, which differ in profitability such that $e_1/h_1 > e_2/h_2$, where e_i is the energy content of a single prey item of type i and h_i is the handling time of type i prey items, and that the predator has a type II (monotonically decelerating) functional response to changes in prey density. The time available for foraging each day can be partitioned into the time devoted to searching for new prey (T_s) and the time devoted to handling prey (T_h), where $T_s + T_h = 1$. This expression assumes that the forager cannot handle prey items at the same time that it searches for new items. Following the same logic as in Holling's (1959) initial derivation of the functional response, the number of prey attacked per unit search time depends on the area searched per unit time (a) multiplied by the prey density (N_i) and the time spent searching (T_s), such that the number of encounters with prey per unit search time = aN_iT_s. Because their spatial distribution is assumed random, either prey 1 or prey 2 individuals might be encountered at any point during search. The time spent handling prey items of type i is $ah_iN_iT_s$, so the total time devoted to handling prey $T_h = aT_s(\beta_1 h_1 N_1 + \beta_2 h_2 N_2)$, where β_i is the probability of attack against each prey item i encountered. The rate of energy gain (or fitness, which we symbolize by w) obtained by feeding on a mixture of the two prey types can be predicted accordingly from the multispecies functional response (Murdoch 1973; Charnov 1976*b*):

$$w = \frac{a(\beta_1 e_1 N_1 + \beta_2 e_2 N_2)}{1 + a(\beta_1 h_1 N_1 + \beta_2 h_2 N_2)} \tag{2.1}$$

It is conceivable that a predator might specialize on either prey type or feed to varying degrees on both prey types. Nonetheless, the optimal solution is for a predator to always attack prey 1 whenever encountered, but conditionally attack prey type 2 (MacArthur and Pianka 1966; Stephens and Krebs 1986). This can be seen by differentiating w with respect to changes in either β_1 or β_2 to determine conditions under which fitness would improve with changes in prey acceptance.

We first see whether fitness would improve by increasing β_1:

$$\frac{\partial w}{\partial \beta_1} = \frac{(e_1 h_2 - e_2 h_1)\beta_2 a^2 N_1 N_2}{(1 + \beta_1 a h_1 N_1 + \beta_2 a h_2 N_2)^2} \tag{2.2}$$

Recall that we earlier assumed that $e_1/h_1 > e_2/h_2$. Since all of the parameters and population variables in equation 2.2 must be positive, $\partial w/\partial \beta_1$ is always positive, signifying that foragers should always accept superior prey whenever encountered ($\beta_1 = 1$). If we do the same for the inferior prey 2,

$$\frac{\partial w}{\partial \beta_2} = \frac{a N_2(e_2 + \beta_1 e_2 a h_1 N_1 - \beta_1 e_1 a h_2 N_1)}{(1 + \beta_1 a h_1 N_1 + \beta_2 a h_2 N_2)^2} \tag{2.3}$$

Fitness would be improved by increasing β_2, that is $\partial w/\partial \beta_2 > 0$, provided that

$$N_1 < \frac{e_2}{a(e_1 h_2 - e_2 h_1)} \tag{2.4}$$

This reasoning indicates that the optimal diet strategy should be an all-or-nothing response: Ignore all poor prey encountered unless the density of better prey falls below a critical threshold, which we symbolize by η_1. This all-or-nothing response implies a step function of the population density of prey species 1.

Phrasing this concept another way, an optimal forager would compare the profitability of prey 2 against the long-term rate of energy gain obtained by specializing solely on prey 1 (Fig. 2.1). This expected long-term rate of gain is calculated by the functional response multiplied by energy content of prey 1 items. A narrow diet would be selectively advantageous when prey 1 individuals are sufficiently common that the expected rate of energy gain obtainable over the long-term by specializing on prey 1 exceeds the short-term gain obtainable by a predator that sneaks a single meal of prey type 2 ($N_1 > \eta_1$). A broad diet would be advantageous if the converse is true ($N_1 < \eta_1$). The decision dictated by the contingency model is a response to the threat of lost opportunities: A predator that always attacks both prey indiscriminately would accordingly waste opportunities for feeding on the more profitable prey (Stephens and Krebs 1986).

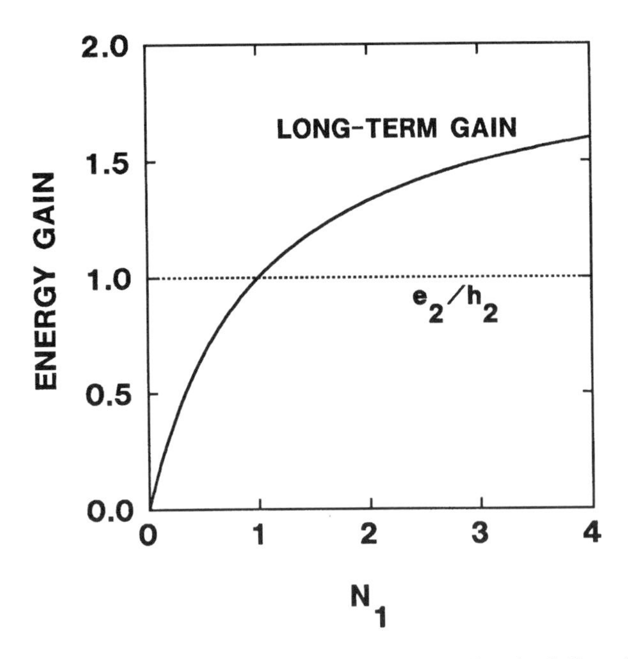

Figure 2.1. Graphical depiction of the contingency model of optimal diet selection. The rate of energy gain for a specialist eating only prey type 1 is monotonically related to N_1, as in the conventional type II functional response. When this average rate of energy gain falls below the instantaneous rate of gain obtainable should a type 2 prey fall into the predators hands (profitability e_2/h_2), then the predator would benefit by facultatively expanding its diet.

2.1.1 Experimental Evidence for Optimal Diets

In the heady days following the initial development of optimal diet choice theory, a number of elegant experimental tests were conducted (see Stephens and Krebs 1986 for a comprehensive review of experiments). By and large, such studies found evidence qualitatively consistent with the general patterns predicted by the basic contingency model, particularly the prediction that diet breadth should narrow with increasing availability of highly profitable prey. Few experiments attempted quantitative (and therefore riskier) predictions of diet selection. Those that did attempt quantitative predictions found that diet inclusion was rarely of the all-or-nothing pattern predicted according to the simplest models (Stephens and Krebs 1986).

This finding is well illustrated by Krebs et al.'s (1977) classic diet choice experiment using great tits (*Parus major*). The birds were first trained to pick mealworms from a conveyor belt as it passed briefly in front of the "forager." Then the experimenters simply manipulated the proportions of mealworms of two sizes placed in randomized order on the conveyor belt and their rate of delivery. Hence, the experimental protocol involved controlled changes in the "density" of most profitable prey to test the prediction that diet expansion should occur when the rate of delivery of preferred prey falls below the critical threshold.

Their results clearly showed that the pattern of acceptance of less profitable prey was indeed consistent with the simple contingency model: Above the critical threshold of 1/20th prey per second, experimental subjects rarely accepted the inferior prey, whereas birds fed indiscriminately when the delivery rate of preferred prey fell below the critical threshold η_1 (Fig. 2.2). The actual pattern of the selection data was less abrupt than the step function strictly predicted by the contingency model. Hence, the behavior was qualitatively consistent with the model, but indicated partial preferences (i.e., imperfect diet selection) rather than pure dietary preferences.

A second type of experiment that can be used to test optimal diet choice theory is changes in selectivity in relation to changes in the proportion of prey available. Such data are commonly used to assess dietary preferences (Manly et al. 1993). There are many statistical measures of feeding preferences and selectivity, of which the simplest is Ivlev's electivity index (I_i/F_i) where I_i = proportion of prey

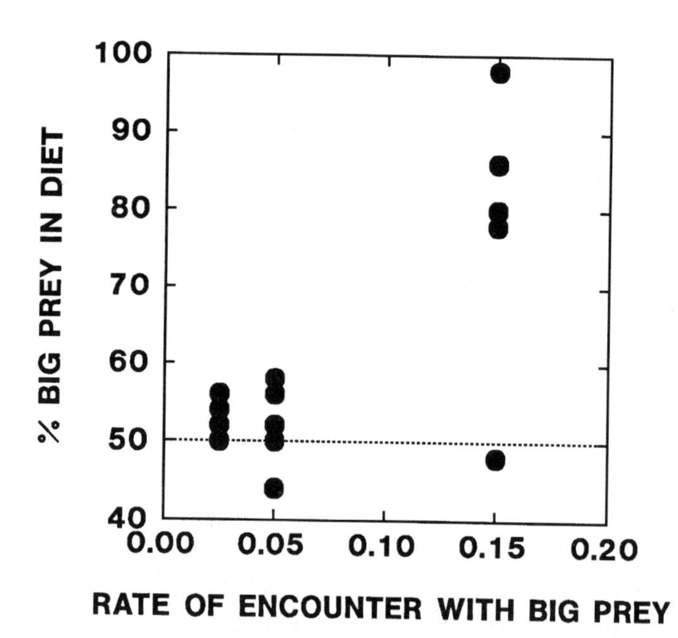

Figure 2.2. Krebs et al.'s (1977) laboratory experiment on diet selection by great tits (*Parus major*). Two sizes of mealworm prey were delivered via a conveyor belt in random order to the forager and the forager's response was recorded. The horizontal axis shows the rate of feeding obtainable by a specialist on the superior prey type. At a threshold rate of delivery of superior prey it benefited birds to expand their diet, hence the percentage selection for superior prey dropped to 50, implying indiscriminate feeding.

type i in the diet and F_i = proportion of prey type i available in environment. Ivlev's index is often used to assess patterns of selectivity when predators are presented with the option of choosing between prey types (Murdoch and Oaten 1975). For a perfectly indiscriminate forager, a scatterplot of I_i versus F_i should fall on a straight line between the origin (when $F = 0$) and the maximum value $I = 1$ (when $F = 1$), provided that the forager perceives and attacks each prey with equal probability. An interesting alternate pattern should occur, however, when predators choose adaptively between prey in a manner consistent with optimal diet choice models (Hubbard et al. 1982).

Assume that total density of prey is held constant but the proportion of prey made available is experimentally manipulated. An adaptive predator should narrow its diet when the proportion of profitable prey (F_1) is such that N_1 exceeds η_1. Hence, one would predict that values of I_1 versus F_i should fall on a straight line for small values of F_1, but jump to a value approaching 1 for large values of F_1 (Hubbard et al. 1982). Partial preferences would tend to blur the sharpness of the discontinuity, but the same qualitative pattern should still occur. Some experimental data are indeed suggestive of such selectivity patterns (Hubbard et al. 1982).

In sum, there is good evidence that the kinds of behavioral decision making predicted by optimality models are at least sometimes displayed by real predators, at least when faced with the relatively simple decisions of experimental environments. There is rather more doubt about the application of such simple models to naturalistic field conditions (Janetos and Cole 1981; Schluter 1981). Field applications of optimal foraging theory probably call for substantially more complex models that take into account variation in the size and spatial distribution of potential prey items as well as variation in motivational state, body condition, and risk proneness of foragers themselves.

2.1.2 Partial Preferences

Some statistical variation would be expected around any biological step function (Stephens 1985), particularly when the function in question is performed by individuals whose motivational state, perceptive abilities, and knowledge about the state of the system are not only likely to be variable but also well outside the measurement capacity of the experimenters (Krebs and McCleery 1984). The existence of pronounced variation in diet choice is scarcely compelling evidence against rate-maximizing behavior and can be readily handled using existing statistical tests (Stephens 1985). Numerous modifications to the basic contingency model of optimal diet selection that might biologically account for partial preferences have been suggested.

The simplest model assumes that the predator is exposed to prey in sequential fashion. In reality, there may often be occasions in which alternate prey are perceived more or less simultaneously by the predator. In such cases, the less profitable prey might occasionally be sufficiently closer than the more profitable prey that the less profitable prey would be attacked under circumstances inconsistent with the strictly sequential contingency model (Waddington 1982). Pulliam (1975) argued that no single food is likely to provide all essential nutritional constituents and that these nutritional needs might constrain diet choice. This process could explain partial preferences, because those foods providing essential nutrients or having lower concentrations of antinutrients such as tannins, must be added to an otherwise energy-maximizing diet (Belovsky 1978; Milton 1979; Lacher et al. 1982; Schmitz et al. 1992; McArthur et al. 1993). We consider nutritional considerations more in Section 2.3 (Balanced nutrient diets).

If animals must sample prey periodically in order to make intelligent diet decisions, then partial preferences should necessarily follow (Krebs et al. 1977; Rechten et al. 1983). This sampling could be required for a predator to reassess profitabilities of potential foods available or perhaps the familiarity gained by the forager via sampling could alter prey encounter rates or handling times (Hughes 1979; Persson 1985). Moreover, if the sequence of encounters with alternate prey is nonrandom (i.e., the predator tends to encounter prey of a given type in bunches), then a sigmoid diet selection curve would result (Lucas 1983; McNamara and Houston 1987).

All of these models are intuitively appealing, at least under some circumstances. In the interests of mathematical simplicity, however, we prefer to simply accept that there are plausible biological circumstances that could explain the widespread occurrence of partial preferences and deal with the ecological ramifications that may result. For this task, we use a simple phenomenological model of partial preferences.

Assume that the cost of deviating from perfect behavioral decision making (i.e., step functions for diet inclusion) is most pronounced when prey population densities differ greatly from the critical threshold η_1. We would therefore expect close correspondence between observed and optimal diet selection when $N_1 << \eta_1$ or when $\eta_1 << N_1$, but rather sloppy correspondence when N_1 is $\approx \eta_1$ (McNamara and Houston 1987). We can model such a situation using:

$$\beta_2 = \frac{e^{z\eta_1}}{e^{z\eta_1} + e^{zN_1}} \tag{2.5}$$

where z is a parameter specifying the magnitude of deviation around the expected step function. The β function has a symmetrical sigmoid shape and is centered around η_1. When z is small, then deviation between the β function and the step function is relatively large, but the two functions converge in the limit as $z \to \infty$ (Fig. 2.3). Similar exponential functions can be directly derived in relation to the energetic cost of deviation (McNamara and Houston 1987) or stochastic control theory (Yoccoz et al. 1993), at the cost of considerably greater computational complexity.

It is not common in behavioral experiments to test feeding preferences under widely varying rates of profitability. Nonetheless, there is fairly solid experimental

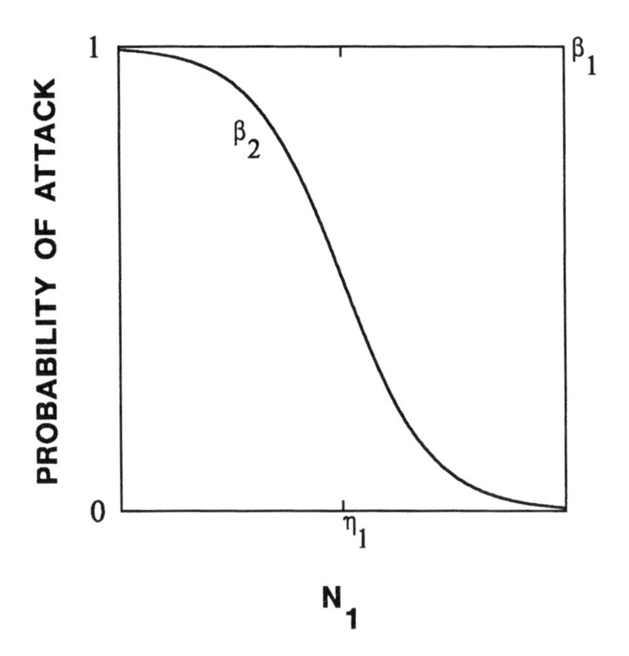

Figure 2.3. Probability of acceptance of the inferior prey type 2 in relation to population density of the superior prey type 1, according to the partial preference model (eq. 2.3).

evidence that predictive ability of contingency models tends to be positively related to the difference in profitability between alternate prey (Johnson and Collier 1989; Osenberg and Mittelbach 1989; Fryxell and Doucet 1993), which is consistent with the β probability.

2.1.3 Diet Selection and the Functional Response

From the point of view of understanding the possible dynamics of interacting populations of predators and prey, the critical linkage is the predator functional response to changes in prey density. If prey are homogeneously distributed in the same habitats, then the rate of consumption by each predator (i.e., the functional response) on each prey type i (X_i) can be calculated by the following formula:

$$X_i = \frac{a\beta_i N_i}{1 + a(\beta_1 h_1 N_1 + \beta_2 h_2 N_2)} \tag{2.6}$$

The rate of consumption of each prey depends on the densities of both prey types (Fig. 2.4). If the density of preferred prey N_1 exceeds the critical threshold η_1, then the rate of consumption is monotonically decelerating, i.e., Holling's

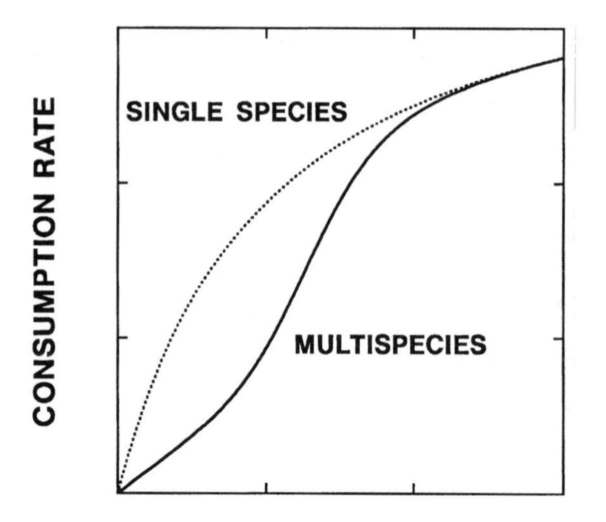

Figure 2.4. Rate of consumption (i.e., the functional response) of preferred prey 1 in relation to density of preferred prey, for a predator using facultative diet selection. When N_1 is large or when N_2 is very small, then the functional response surface is monotonically decelerating, like a classic type II functional response. When N_1 is small and N_2 is abundant, then facultative diet expansion leads to an inflection in the functional response, implying positive density dependence.

(1959) type II response is recovered. On the other hand, if $N_1 << \eta_1$, then each prey is consumed proportionate to its abundance. Consequently, when preferred prey are scarce and alternate prey are abundant, the rate of consumption of preferred prey is diluted by alternate prey. This result stems from the adaptive predator devoting a greater proportion of its time to handling alternate prey and commensurately less time to search. When alternate prey are not common or when the density of preferred prey exceeds the critical threshold η_1, however, the dilution of predation risk is minuscule. All of this is simply another way of saying that facultative diet selection produces an inflection in the functional response (Holt 1983; Gleeson and Wilson 1986; Persson and Diehl 1990; Fryxell and Lundberg 1994).

Per capita rates of predation mortality are affected accordingly by diet choice. As outlined in Chapter 1, a monotonically decelerating functional response implies that the per capita risk of mortality due to predation declines at a decelerating rate with prey density. For systems in which a predator exercises adaptive diet selection, per capita risk for preferred prey is negatively related to density either when $N_1 << \eta_1$ or when $N_1 >> \eta_1$. The inflection in the functional response when $N_1 \approx \eta_1$ produces an increase in the per capita risk of mortality due to predation.

2.1.4 Experimental Tests of the Functional Response

Experimental tests of functional responses when predators are presented with a single prey type are quite common in the literature. It is therefore all the more remarkable how few functional response experiments have been conducted with more than one prey size class or with more than one prey species available. Even when there is no particular indication that predators are exercising behavioral selectivity in choosing their diet, multiprey functional responses provide unique insights into the complex ways that rates of consumption change when the resource base is altered (Ranta and Nuutinen 1985; Colton 1987). This exercise can be sobering when the ultimate objective is to apply functional response parameters in models of trophic interactions.

There is some indication, nonetheless, that adaptive diet selection can induce inflections in the multispecies functional response, which implies that the per capita risk of prey mortality increases with increasing prey density. For example, beavers in boreal forests of North America are commonly faced with several species of woody plants of varying energetic profitability. Fryxell and Doucet (1993) recreated an artificial forest of 3 woody plant species, trembling aspen, speckled alder, and red maple, in an enclosure inhabited by wild-caught beavers. Experimental estimates of energy content and handling time indicated that aspens were more profitable than alders, which in turn were more profitable than red maples. Initially, the forest was composed of equal densities of the 3 species; thereafter sapling removals by the beavers skewed the relative proportions of the remaining plant species. As predicted by the contingency model, beavers initially cut down mostly aspens, then shifted to mostly alders at intermediate stages of depletion, and were finally forced to forage on maple saplings. The functional responses recorded in these multispecies experiments differed considerably from single-species functional responses (Fig. 2.5). More importantly, there was evidence of an inflection in the multispecies functional response at low sapling densities, which produced density-dependent mortality at low sapling densities (Fig. 2.5). This is perhaps one of the better controlled examples of facultative diet choice and the functional response, but there are many experiments on switching behavior leading to sigmoid functional responses by predators (Lawton et al. 1974; Murdoch and Oaten 1975; Akre and Johnson 1979). Such data sets are circumstantially consistent with optimal diet selection models. Since no data are available on profitability, however, it is impossible in many cases to ascertain whether the switches were related to energy-maximizing behavior. There are well-documented examples of switching, however, that are unrelated to energy-maximizing diet choice (Hughes and Croy 1993). Other mechanisms, such as predator search images, spatial segregation of different prey in the environment,

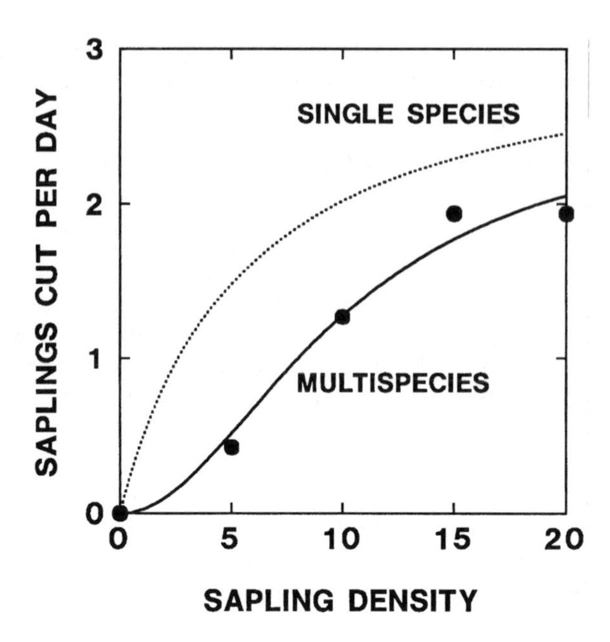

Figure 2.5. Observed functional responses of beavers in a large enclosure presented with a single species of highly profitable plant prey (dotted line) or a mixture of profitable and unprofitable plant prey (solid line) (data from Fryxell and Doucet 1993). At low prey densities beavers expanded their diet to include substantial quantities of poorer quality prey, apparently inducing an inflection in the functional response.

risk avoidance by foragers, or optimal patterns of foraging effort could also cause sigmoid functional responses.

Facultative expansion of the diet as N_1 falls below the critical threshold η_1 should lead to rapid dilution of predation risk to the preferred prey. Partial preferences will tend to expand the region of positive density dependence. In either case, the positive density dependence resulting from adaptive diet expansion could potentially tend to counteract or compensate for perturbations away from population equilibria (Gleeson and Wilson 1986; Abrams 1987 *a,b;* Persson and Diehl 1990; Fryxell and Lundberg 1994). Whether it does so or not depends on other population parameters.

2.1.5 Diet Choice and Population Dynamics

In order to understand the implications of diet selection for trophic dynamics, we embedded our simple models of diet choice behavior in a modified form of the Lotka–Volterra equations, such that density-dependent recruitment serves to regulate prey in the absence of predation:

$$\frac{dN_1}{dt} = r_1N_1\left(1 - \frac{N_1}{K_1}\right) - X_1P \tag{2.7}$$

$$\frac{dN_2}{dt} = r_2N_2\left(1 - \frac{N_2}{K_2}\right) - X_2P \tag{2.8}$$

$$\frac{dP}{dt} = cP(e_1X_1 + e_2X_2) - dP \tag{2.9}$$

where r_i is the maximum per capita recruitment rate of prey *i*, K_i is the carrying capacity of prey *i*, *c* is a parameter converting the rate of prey consumption into the per capita rate of predator reproduction and *d* is the per capita mortality rate of the predators. We compared the time dynamics of these models incorporating diet choice by predators with that of traditional predator–prey models (Chapter 1), in which predators only feed on a single prey. In all cases, of course, we used identical parameter combinations.

Depending on the quality of alternate prey, stability can be either diminished or enhanced by adaptive diet choice (Fryxell and Lundberg 1994). When alternate prey are more than sufficiently profitable for predator subsistence ($e_2/h_2 > d/c$), then the proportion of stable parameter combinations is generally diminished (Fig. 2.6*A*). In contrast, stability is often enhanced when alternate prey are marginally sufficient for predator persistence ($e_2/h_2 \approx d/c$) (Fig. 2.6*B*).

The difference in stability properties of diet selection depends on population equilibria in relation to the point of diet expansion. Whenever the equilibrium density of preferred prey in the system (N_1) is close to the point of diet expansion (η_1), then the density dependence due to the inflection in the functional response is sufficient to allow stabilization (Fryxell and Lundberg 1994). Such is the case when $e_2/h_2 \approx d/c$, because a predator feeding exclusively on a superabundant supply of poor quality prey 2 would acquire just enough energy to allow $dP/dt \approx 0$. By definition, the rate of energy gain when feeding on high quality prey at a density of η_1 is equivalent to e_2/h_2, therefore $dP/dt \approx 0$ also for a predator feeding on high quality prey near the point of diet expansion. Hence, an adaptive predator can drive preferred prey to low enough density to cause expansion of its diet. Diet expansion does not lead to further population growth of predators, so the rates of change of all components slow down and the system approaches equilibrium.

In contrast, when $e_2/h_2 >> d/c$, then the predator's rate of change is positive near the point of diet expansion, hence the system does not stabilize, but predators continue to drive both high quality and poorer quality prey to even lower densities, which are usually outside the range of density-dependent mortality. Hence, the system is not stabilized, even though there is a density-dependent prey response. Density dependence is essential for the natural regulation of a population (Sinclair 1989; Murdoch 1994), but density dependence of one component of the commu-

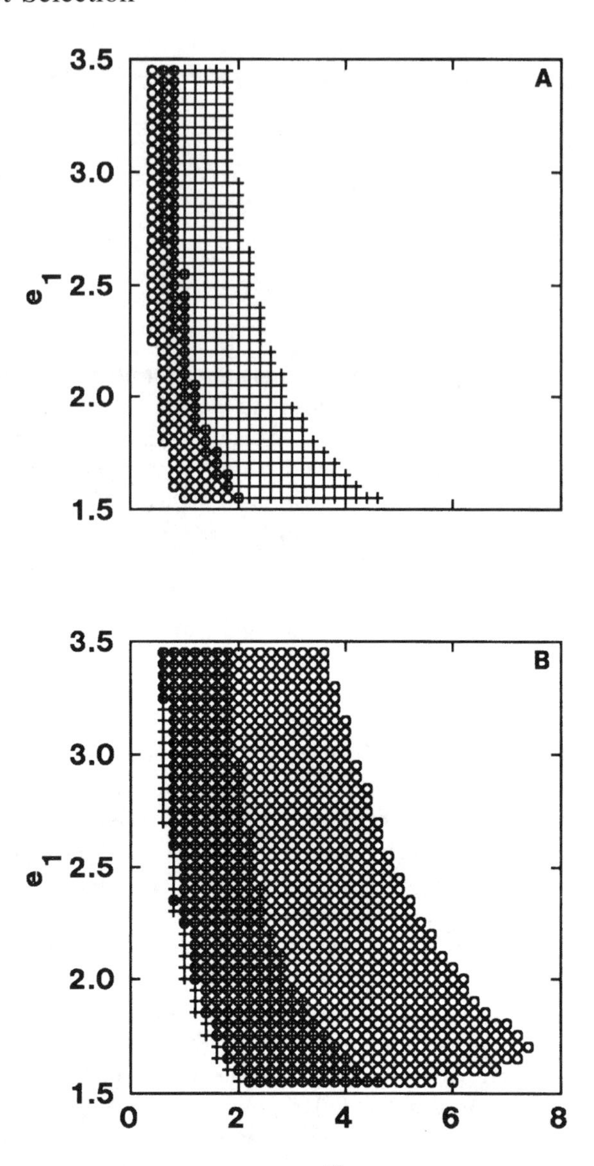

Figure 2.6. Stable combinations of the e_1 versus K_1 parameter space for predator–prey interactions in which the predator either behaves as a specialist on the superior prey (crosses) or facultatively expands its diet to include the poorer prey type 2 (circles) when $N_1 < \eta_1$. Two scenarios are depicted: (*A*) alternate prey are profitable enough to allow predator increase, $e_2/h_2 = 1.2$; (*B*) alternate prey are too unprofitable to permit predator persistence, $e_2/h_2 = 0.8$. Other parameter values were as follows: $a = c = d = h_1 = h_2 = r_1 = r_2 = 1$, $z = 4$, $K_2 = K_1$.

nity may be insufficient to guarantee a stable community. Facultative diet selection leads to weak density dependence that is often of insufficient magnitude to stabilize the community.

The effect of facultative selection can be readily seen in the zero growth isocline for preferred prey (combinations of N_1, N_2, and P at which $dN_1/dt = 0$), as shown in Fig. 2.7. Density dependence is displayed by the parts of this surface that have a negative slope in relation to changes in N_1 (i.e., $\partial(dN_1/dt)/\partial N_1 < 0$). Only a small portion of the prey 1 zero isocline has a negative slope, so density dependence comes into play only over a small range of conditions (Fig. 2.7).

Facultative diet selection is analogous in some ways to classic models of predator "switching" behavior. Both behavioral mechanisms induce density-dependent effects on prey mortality, which is sometimes sufficient to stabilize community interactions (Murdoch and Oaten 1975; Comins and Hassell 1976; Tansky 1978; Gleeson and Wilson 1986; Persson and Diehl 1990; Fryxell and Lundberg 1994). Unlike switching behavior, the stabilizing effects of facultative diet selection should often be selectively advantageous at the individual level (Holt 1983; Fryxell and Lundberg 1994, but see section 2.2). A similar stabilizing effect has been demonstrated for Nicholson–Bailey models incorporating alternate host genotypes of differential attractiveness to a facultative parasitoid

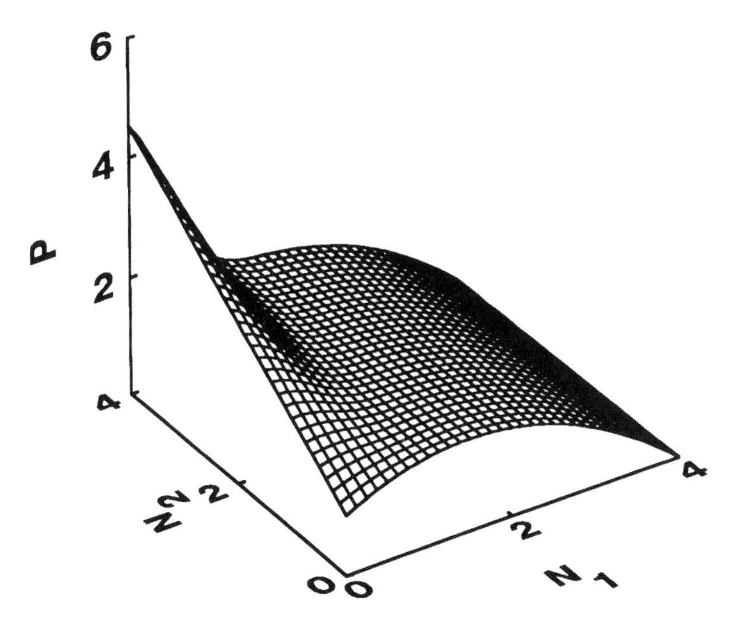

Figure 2.7. Zero growth isocline for a system with a predator exhibiting facultative diet selection between two prey with different profitabilities. Parameter values were as follows: $a = c = d = h_1 = h_2 = r_1 = r_2 = 1$, $z = K_2 = K_1 = 4$, $e_1 = 2$, $e_2 = 0.8$.

(Mangel and Roitberg 1992), suggesting that the diet choice paradigm may prove to be stabilizing in different community contexts.

Pronounced partial preferences enhance the stabilizing effect of diet selection (Fryxell and Lundberg 1994; van Baalen 1994). This stabilizing effect is due to expansion of the range of density-dependent mortality for prey. When predators expand their diet slowly with changes in preferred prey, then this implies a broader range of density-dependent responses. Hence, the kind of behavioral "sloppiness" or "error" around perfectly optimal decisions commonly observed in experimental settings could by itself have a profound influence on system dynamics, suggesting that deeper understanding of the processes causing variability around optimal behavioral solutions may be just as important in understanding trophic implications as the optimal decisions themselves. Fortunately, there are signs that behavioral ecologists are beginning to come to grips with the thorny problem of ascribing, and more importantly predicting, the biological causes of partial preferences (McNamara and Houston 1987; Yoccoz et al. 1993).

Diet selection also tends to bound the amplitude of population fluctuations in unstable systems (Gleeson and Wilson 1986; van Baalen 1994; Krivan 1996), because the lower "hump" in the prey isosurface (Fig. 2.7) sometimes traps cyclical population trajectories, avoiding the perilously low population densities that would otherwise occur. This is an important community property, because small populations, such as those at the troughs of a cycle, are at risk of genetic bottlenecks and chance extinction via demographic stochasticity (Leigh 1981; Goodman 1987; Pimm et al. 1988; Lande 1993). A reduction in the amplitude of population fluctuations should therefore improve the prospect of community persistence, which in the long term may be a more important property than stability per se for the conservation of biotic diversity.

2.2 EVOLUTIONARY DYNAMICS OF DIET SELECTION

Thus far, we have confined our attention to the effects of adaptive foraging behavior on trophic interactions in the community. It is at least as interesting to turn this around and ask what the effect of community dynamics might be on the evolution of alternate behavioral strategies in the predator population. There are many ways one could approach this sort of problem. For example, one could postulate a specific genetic relationship and model changes in allele frequencies over time. Foraging behavior probably comprises such a complex combination of physiological and behavioral traits, however, that such a population genetic model is somewhat unwieldy (Fryxell 1997). Here we pursue a different tack, assuming that genetic mechanisms exist for four alternate genotypes: perfect specialist behavior (feeding only on prey 1 or prey 2), perfect generalist behavior (feeding indiscriminately on both prey types), and facultative diet choice behavior (feeding as a specialist when $N_1 > \eta_1$, but as a generalist when $N_1 < \eta_1$). We then test whether one behavior is always an evolutionarily stable strategy (ESS) or whether there are circumstances leading to coexistence of alternate genotypes within the same population (a multiple ESS).

Attributes other than feeding behavior were assumed constant across genotypes. We then simulated the temporal dynamics of a system with the following structure:

$$\frac{dN_1}{dt} = r_1 N_1 \left(1 - \frac{N_1}{K_1}\right) - \sum_{j=1}^{4} X_{1j} P_j \tag{2.10}$$

$$\frac{dN_2}{dt} = r_2 N_2 \left(1 - \frac{N_2}{K_2}\right) - \sum_{j=1}^{4} X_{2j} P_j \tag{2.11}$$

$$\frac{dP_j}{dt} = cP_j(e_1 X_{1j} + e_2 X_{2j}) - dP_j \tag{2.12}$$

where $\beta_1 = 1$ and $\beta_2 = 0$ for a predator specializing on prey 1 ($j = 1$), $\beta_1 = 0$ and $\beta_2 = 1$ for a predator specializing on prey 2 ($j = 2$), $\beta_1 = 1$ and $\beta_2 = 1$ for a generalist predator ($j = 3$), and $\beta_1 = 1$ and β_2 was calculated using equation 2.5 for the facultative predator ($j = 4$). We simulated dynamics of these systems over a range of parameters to test whether ESS strategies depended on environmental conditions.

The first surprise is that the facultative strategy does not always displace other predator genotypes: There are parameter combinations in which either a predator specializing on prey 1 or the generalist genotype is an ESS and the other genotypes cannot compete effectively over time (Fig. 2.8). The second surprise is that some parameter combinations select for a multiple ESS, comprising the facultative genotype and either the predator specializing on prey 1 or the generalist genotype (Fig. 2.8). Hence, flexible diet selection would not always be evolutionarily stable: It can often be invaded successfully by individuals that do not expand their diet when prey change in relative abundance, in some cases driving facultative predators extinct.

The explanation can be traced back to (1) the effect of population fluctuations over time on the adaptive landscape and (2) the slightly suboptimal effects imposed by partial preferences. Recall that the rate of energy gain of a specialist on prey 1 is higher than that of a generalist when $N_1 > \eta_1$, but the converse is true when $N_1 < \eta_1$. A facultative predator with partial preferences has equal fitness to a specialist on prey 1 when $N_1 >> \eta_1$ and equal fitness to a generalist when $N_1 << \eta_1$, but slightly lower fitness when $N_1 \approx \eta_1$.

If dynamics of the system are such that most of the time preferred prey densities are well below this fitness threshold, then generalists are perpetuated, but specialists and facultative predators gradually disappear (Fig. 2.9*A*). In contrast, if most of the time the system behaves such that N_1 is well above the isofitness threshold, then

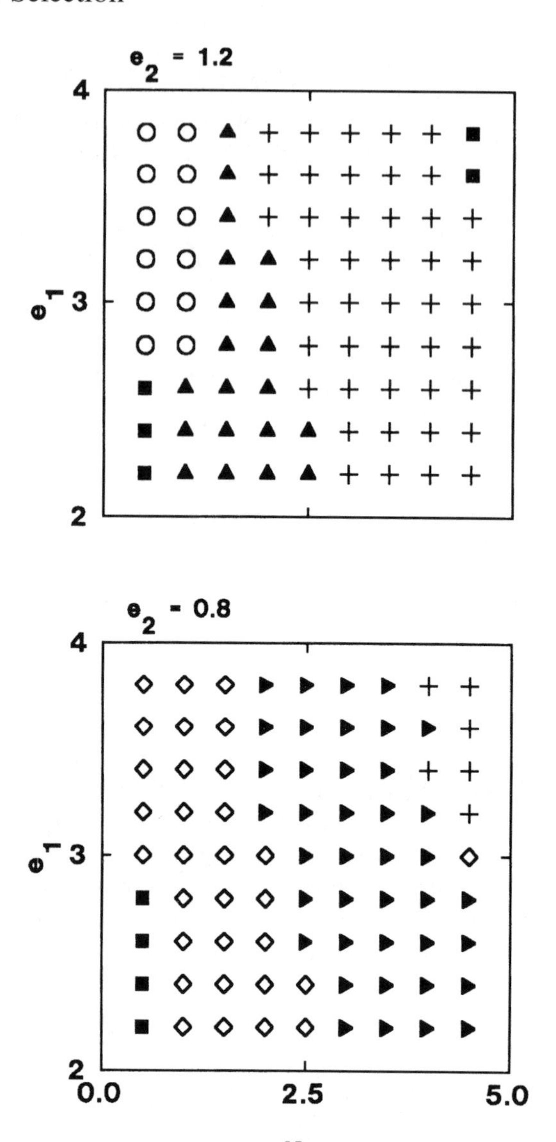

Figure 2.8. Evolutionarily persistent outcomes of dynamic competition among specialists, generalists, and facultative predators over a range of values of the z and K parameters. + implies that facultative predators win, ▲ implies that facultative predators coexist with generalists, ○ implies that generalists win, ▶ implies that facultative predators coexist with specialists, ◇ implies that specialists win, and ■ implies that all genotypes go extinct. Parameter values were as follows: $a = c = d = h_1 = h_2 = r_1 = r_2 = 1$, $K_2 = K_1$. In top figure, $e_2 = 1.2$, whereas in bottom figure $e_2 = 0.8$. Simulations were conducted for 500 time steps and the criterion for persistence was that population densities of each genotype > 0.001.

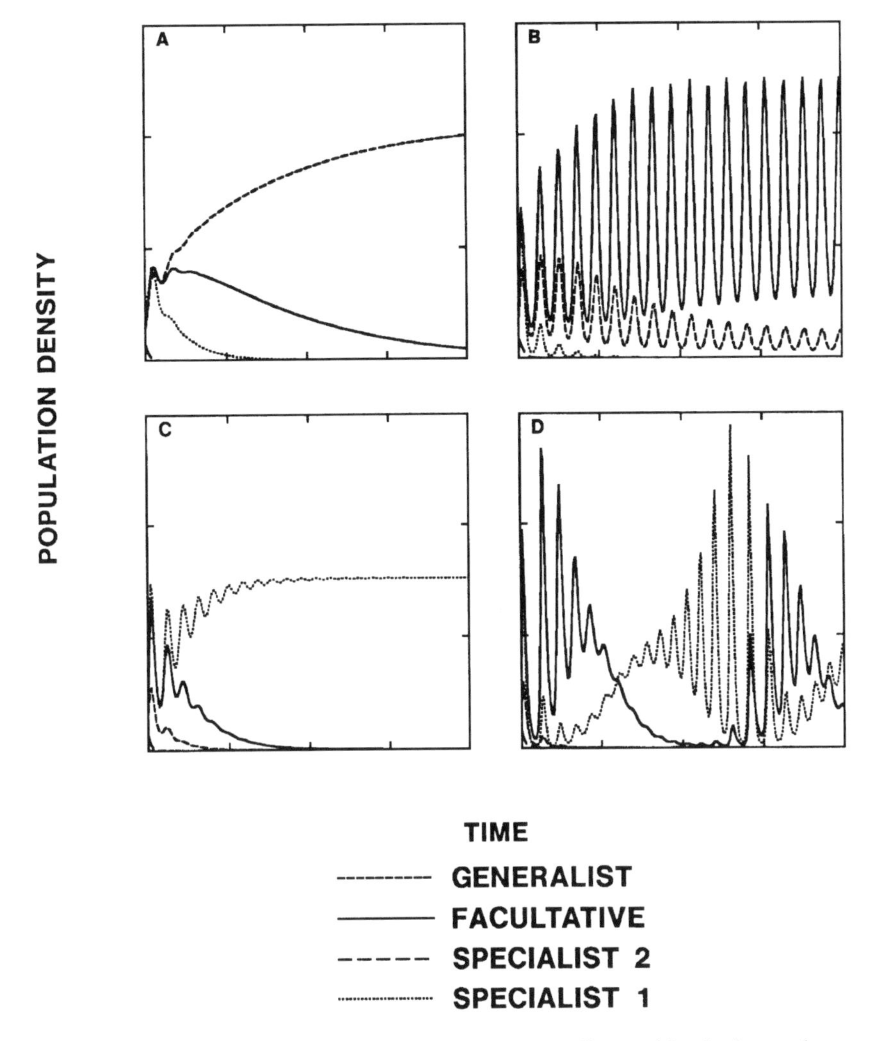

Figure 2.9. Time dynamics of competing specialists, generalists, and facultative predators, for a range of parameter combinations spanning the range shown in Figure 2.8 (*A*: $K_i = 1.0$, $e_1 = 3.0$ and $e_2 = 1.2$; *B*: $K_i = 2$, $e_1 = 3.0$ and $e_2 = 1.2$; *C*: $K_i = 1.5$, $e_1 = 3.2$ and $e_2 = 0.8$; *D*: $K_i = 3.0$, $e_1 = 3.2$ and $e_2 = 0.8$). Other parameter values were as follows: $a = c = d = h_1 = h_2 = r_1 = r_2 = 1$, $z = 4$.

specialists on prey 1 are perpetuated, but generalists and facultative predators gradually disappear from the gene pool (Fig. 2.9*C*). If cyclical fluctuations dominate such that N_1 fluctuates with equal probability above and below the isofitness threshold, then this allows dominance by the facultative genotype (Fig. 2.9*B*).

Many parameter combinations, however, permit coexisting strategies: either the facultative and the specialist on prey 1 genotypes or the facultative and the generalist genotypes (Fig. 2.8). In such systems, the pure genotype would have a competitive advantage over the facultative genotype at equilibrium. This competitive advantage leads, however, to population cycles of increasing amplitude for the pure genotype, paving the way for reciprocal domination by the facultative genotype (Fig. 2.9*D*). As density-dependent predation by the facultative genotype decreases the amplitude of population fluctuations, the pure genotype dominates once again. In sum, the facultative genotype prospers during periods of pronounced community instability, whereas the pure genotypes prosper during periods of relative calm.

Such dynamics are not chaotic, in that there is a recurrent periodicity, but rather protracted limit cycles. Evolutionary dominance is never assured in such systems, because ascendency by a given behavioral genotype inevitably changes the dynamical characteristics to favor another genotype. In a sense, some genotypes are a victim of their own success, whereas some genotypes depend for their own persistence on the instability of others. Similar patterns have been observed in models with competition between diet-balancing predators (Abrams and Shen 1989).

Although we have confined our attention to purely deterministic systems, environmental stochasticity should also influence evolutionary dynamics in analogous fashion (Wilson and Yoshimura 1994). For example, if prey populations responded to stochastic variation in weather conditions, then this could alter ESS outcomes depending on the resulting frequency distribution of prey population densities over time. For intrinsically unstable systems, however, the deterministic temporal signal would probably overwhelm such stochastic variability. Our conclusion is that evolutionary outcomes, at least in this context, could be dependent on temporal dynamics of the community, specifically in relation to optimal behavioral thresholds.

Such mechanisms could play a role in maintaining variability in foraging behavior within populations (West 1988; Ritchie 1988; Burrows and Hughes 1991; Skulason et al. 1993; Skulason and Smith 1995). Populations subject to large fluctuations in abundance over time may not have consistent enough selection to lead to a single predominant genotype. There are, of course, many other mechanisms contributing to the maintenance of genetic variability.

2.3 BALANCED NUTRIENT DIETS

The energy-maximizing diet model should be most appropriate to organisms that must choose among prey types that differ according to a single nutritional parameter (i.e., protein or energy content) or handling time, but that are otherwise similar in chemical composition. This assumption might be fine for many carnivores, for example, whose animal prey often have a similar mix of carbohydrates, fats, amino acids, and the like. It is less obvious whether it should necessarily be the case for herbivores, faced with an array of plant foods that broadly differ with respect to chemical composition. Accordingly, there has been a persistent alternative view among some behavioral ecologists that foragers may need to choose diets that maintain a balanced ratio of nutrients, rather than trying to maximize intake of energy or any other single nutrient (Westoby 1974, 1978; Pulliam 1975; Rapport 1980; Raubenheimer and Simpson 1993). Incidentally, it matters little for our purposes whether the forager is balancing intake of beneficial constituents (e.g., amino acids, carbohydrates, or protein) or harmful constituents (e.g., toxins, secondary compounds, or structural compounds). The crucial point is that a balanced diet is more beneficial than an overload of any single constituent.

Another way to explain the balanced nutrient hypothesis is that forager fitness could be dictated by intake of the most limiting nutrient (Tilman 1982; Abrams 1987*a*). Hence, a contour plot of predator fitness in relation to intake of two hypothetical prey types should produce a series of L-shaped isofitness curves (Fig. 2.10). As a consequence, increased consumption of one resource beyond the optimal ratio (the apex of the L) would not improve fitness, because another resource would then simply become limiting (Tilman 1982; Abrams 1987*a*; Abrams and Shen 1989).

We term this process the balanced nutrient hypothesis, although identical

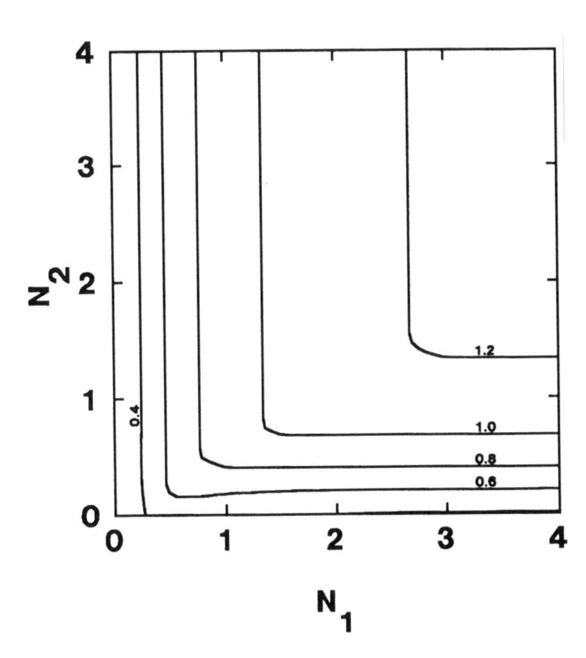

Figure 2.10. Contours of equal fitness for a balanced nutrient model in relation to densities of prey species 1 and 2. Lines connect combinations of prey at which a facultative predator has equal fitness. Fitness is highest for contours farthest from the axes.

processes are also known as "nutritional wisdom" (Westoby 1978), "perfectly complementary resources" (Leon and Tumpson 1975), "essential nutrients" (Tilman 1982), or "nonsubstitutable resources" (Abrams 1987*a*; Abrams and Shen 1989). In each case the process is identical: A forager should allocate its foraging activities such that it maintains constant proportions of several incoming nutrients in its diet. If these nutrients occur in different fractions among different prey types, then this implies that prey types in the diet will also occur in a constant ratio, which we symbolize by the parameter ω. Restricting ourselves to two prey types, for simplicity, then the balanced nutrient hypothesis predicts that rates of prey consumption will be balanced such that $X_1/X_2 = \omega$.

If prey are homogeneously mixed in the environment, then a given period of search will turn up a random sample of both prey types, each of which can either be accepted with a probability β_i or discarded. One would also assume that there is ordinarily a time expenditure in handling each prey item that is accepted. We assume for simplicity that there is a negligible time cost of rejecting prey items that are not accepted. Hence, the optimal ratio ω of prey can only be maintained if the ratio of the probabilities of acceptance also changes with relative availability in the environment, scaled to the following rule:

$$\omega = \frac{\beta_1 N_1}{\beta_2 N_2} \tag{2.13}$$

This rule implies that if the ratio of prey densities in the environment (N_1/N_2) falls below the optimum ratio (ω), then every prey 1 item that is encountered ought to be taken ($\beta_1 = 1$), whereas a lesser proportion of prey 2 items ought to be taken ($\beta_2 = N_1/\omega N_2$). Conversely, if the ratio of prey in the environment exceeds the optimum ($N_1/N_2 > \omega$), then all prey 2 items encountered ought to be accepted ($\beta_2 = 1$), whereas only a fraction of those of prey 1 should be accepted ($\beta_1 = \omega N_2/N_1$). As in the simple contingency model, diet preferences should be mediated by prey availability (Fig. 2.11). In contrast to the simple contingency model, however, the probability of acceptance should depend on densities of both prey types. The pattern of selectivity should also be the reverse of a nutrient-maximizing forager, in that selectivity should be inversely related to relative availability (Abrams 1987*a*). Moreover, the optimal strategy is no longer an all-or-nothing response, but rather partial preferences for one prey or the other, with partial preferences reversing according to changes in prey relative abundance in the environment.

One can model the ecological implications of a balanced diet premise by inserting the partial preferences β_i into the standard multispecies functional response equation (eqn. 2.4):

$$X_1 = \frac{aN_1}{1 + a(h_1 + h_2/\omega)N_1} \quad \mathit{if} \quad \frac{N_1}{N_2} < \omega \tag{2.14}$$

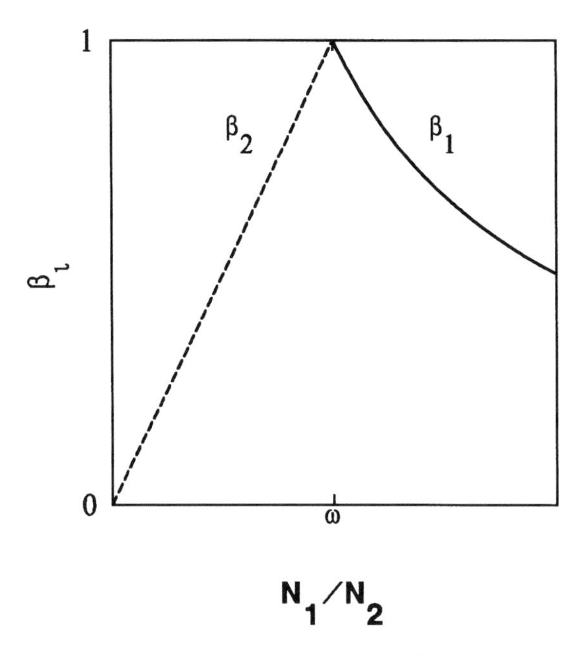

Figure 2.11. Probability of acceptance of prey type 1 (β_1) and prey type 2 (β_2) for a predator with balanced nutrient requirements, in relation to the ratio of prey densities.

$$X_1 = \frac{a\omega N_2}{1 + a(\omega h_1 + h_2)N_2} \quad \textit{if} \quad \frac{N_1}{N_2} > \omega \tag{2.15}$$

$$X_2 = \frac{aN_1/\omega}{1 + a(h_1 + h_2/\omega)N_1} \quad \textit{if} \quad \frac{N_1}{N_2} < \omega \tag{2.16}$$

$$X_2 = \frac{aN_2}{1 + a(\omega h_1 + h_2)N_2} \quad \textit{if} \quad \frac{N_1}{N_2} > \omega \tag{2.17}$$

When the abundance of alternate prey species 2 is pronounced relative to that of a focal prey species 1, then $N_1/N_2 < \omega$ and the functional response surface (X_1) is qualitatively similar to the classic type II response (Fig. 2.12). It is not quite identical, because the diluting effect of prey 2 becomes ever more pronounced the smaller that N_1 is relative to N_2. The curious prediction of the balanced diet hypothesis occurs when $N_1/N_2 > \omega$. Under these circumstances, in which intake of prey 2 limits predator fitness, then the probability of acceptance of prey 1 and therefore consumption of prey 1 should decline relative to the rate when N_2 is high (Abrams 1987*a*). That is why the consumption rate on species 1 can sometimes be dictated entirely by the density of species 2, and the converse for species 2. In summary, the rate of intake of a facultative forager should be similar to that of

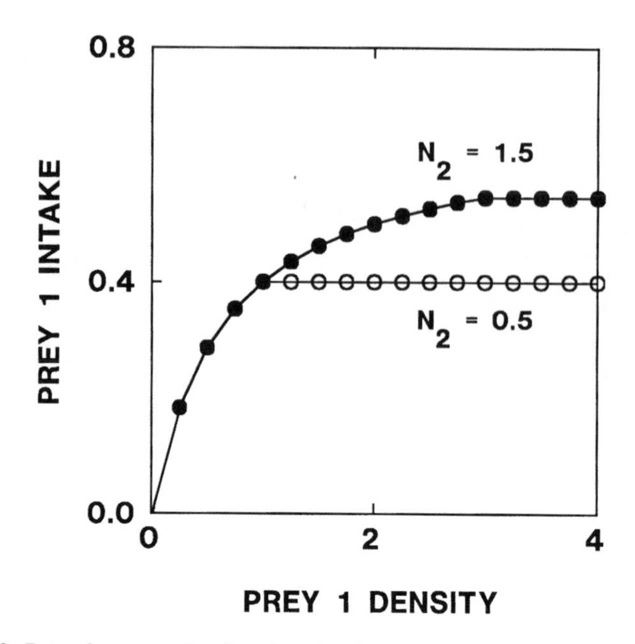

Figure 2.12. Rate of consumption (i.e., functional response) of prey type 1 for a facultative predator with balanced nutrient requirements in relation to densities of both prey types. Consumption of prey 1 declines when N_2 is small, even though N_1 is large, because the predator actually suffers loss of fitness if it consumes prey type 1 in proportion to availability.

a specialist forager when $N_1/N_2 \ll \omega$, similar to a generalist forager when $N_1/N_2 < \omega$, and unlike either (in that X_1 increases with N_2) when $N_1/N_2 > \omega$.

2.3.1 Experimental Evidence for Balanced Diets

These patterns suggest three critical observations that should discriminate between the balanced nutrient and the energy-maximization hypotheses. First, the proportion of each prey type in a balanced diet should remain roughly constant as resource availability changes, particularly when the ratio of prey densities is manipulated. In contrast, energy-maximizing models suggest that the proportion of each prey species in the diet should reflect its ratio in the environment when densities fall below a critical threshold, but become highly skewed when the density of preferred prey is high. Second, the fitness of animals fed a mixed diet should be greater than that of experimental animals fed a monospecific diet. Third, the model for balanced nutrient intake predicts that rates of consumption of each prey type should be enhanced by increased density of alternate prey, whereas the contingency model of diet selection predicts that adding alternate prey would either have no effect or negative impact on the rate of consumption of preferred prey, depending on the optimal diet.

There is ample evidence that nutrient balancing is important in many species. For example, Kitting (1980) found that limpets tended to maintain constant proportions of different algal species in its diet, in spite of substantial changes in prey abundance over both time and space. Milton (1979) found that howler monkeys selected far more leafy matter than would be expected on the basis of energy-maximizing models but consistent with the balanced diet hypothesis. Newman et al. (1992) tested whether sheep reject previously accepted items in order to maintain a balanced diet. They showed that sheep previously fed grass were likely to choose clover than grass in a subsequent trial, whereas the converse was true for sheep previously fed clover. This was surprising, because sheep do better feeding exclusively on clover. Dearing and Schall (1992) found dramatic spatial variability in the diet of the herbivorous whiptail lizard (*Cnemidophorus murinus*), but found that the total intake of nutrients was similar in spite of these differences in diet. Perhaps most importantly, there was little indication that any single nutritional parameter adequately predicted diet composition. Energy-maximizing models have often proved inadequate in explaining the mix of species in herbivore diets (Belovsky 1978, 1986; Milton 1979; West 1988; Newman et al. 1992; Doucet and Fryxell 1993; Penning et al. 1993; Owen-Smith 1993, 1994), hinting that balancing nutrients may be particularly important for herbivores.

Several studies have shown that animals on mixed ad libitum diets show enhanced rates of growth relative to control animals on pure diets (Rapport 1980; West 1988; Penning et al. 1993; Raubenhaimer and Simpson 1993; Bernays et al. 1994). The studies by Penning et al. (1993) on a large marine gastropod, *Dolabella auricularia,* show in particularly convincing fashion that these herbivores grow almost an order of magnitude faster when raised on a mixture of algal foods, than when raised on a monospecific diet. Moreover, the gastropods clearly showed a preference for the rarer algal species, when presented with highly skewed proportions of algal species. They also showed a tendency to switch algal preferences, after forced feeding on a single species.

Hence, there is ample evidence of diet patterns that are qualitatively consistent with the balanced nutrient hypothesis, particularly in herbivores. Indeed, if the recent literature is any guide, there is far more field evidence consistent with the balanced nutrient hypothesis than the nutrient-maximizing hypothesis. Curiously, this predictive power could reflect a lack of quantitative a priori predictions. To our knowledge, there is no theoretical basis for predicting which nutrients ought to be balanced and in which proportions. Hence, only qualitative experimental tests are possible, which are always going to be harder to reject than explicit numerical predictions. A second cautionary note concerns the quality of parameters used in many experimental tests of nutrient maximizing models. Rigorous tests require detailed knowledge about both nutrient content (which is frequently available) and handling time (which is rarely available), particularly for herbivores, for which processing time in the gut is arguably the major component of handling time. When interspecific variation in handling time is included, energy-maximizing models tend to have much better predictive power than when handling time is presumed constant and only nutrient content varies (Belovsky 1978, 1986; Doucet and Fryxell 1993). There is some evidence, moreover, that handling time itself changes seasonally, in accordance with shifting nutritional stresses and internal state of herbivores (Owen-Smith 1994).

It is less well established whether the presence of an alternate prey species can actually enhance the rate of consumption of any single prey species in the ensemble (Abrams 1987*a*). There are some tantalizing hints in the literature, however, that this conjecture might be true. Rapport (1980) showed that presenting a second algal species to the generalist ciliate predator *Stentor coerulus* enhanced the rate of consumption of another algal species, although this pattern was not replicated when a protozoan was presented as the alternate prey species. Not surprisingly, fitness was higher for *Stentor* fed a mixed algal diet than those fed monospecific diets, presumably due to complementarity in nutrient composition. Similarly, Kitting's (1980) experiments with limpets suggest that intake was enhanced, rather than diminished, by the presence of alternate prey, although in this case no data were presented regarding nutrient composition or the fitness of predators on mixed versus pure diets.

2.3.2 Balanced Diets and Population Dynamics

Despite the belief that diet balancing or resource complementarity might be relatively common in nature, its dynamic implications have received relatively little attention by population ecologists. The greatest amount of work has been devoted to competition when critical resources are complementary, which is thought to be particularly relevant for many plants (Leon and Tumpson 1975; Tilman 1982). In keeping with our earlier theme, we concentrate on adaptive diet balancing by herbivores or other predators, which has received far less attention (Abrams and Shen 1989). One can model the dynamics of such a predator–prey interaction by substituting the balanced diet functional responses shown in equations 2.14–2.17 into the predator–prey model shown in equations 2.7–2.9.

The balanced diet model assumes that the predator continually feeds on both prey, to varying degrees. We therefore compared the dynamics of the balanced diet model to a model with a pure generalist predator (both β_1 and $\beta_2 = 1$ all the time). Our results suggest that diet balancing tended to have trivial effects on the degree of stability (Fig. 2.13). It is true that diet balancing affected which parameter combinations were stable, but by and large the proportion of stable parameter combinations was very similar to that of a purely generalist system. When the energy content of alternate prey was higher than subsistence ($e_2/h_2 > d/c$), then some parameter combinations that are unstable for a pure generalist predator become stable. But diet balancing also leads to compensatory changes in the range of feasible parameter combinations. That is, some parameters

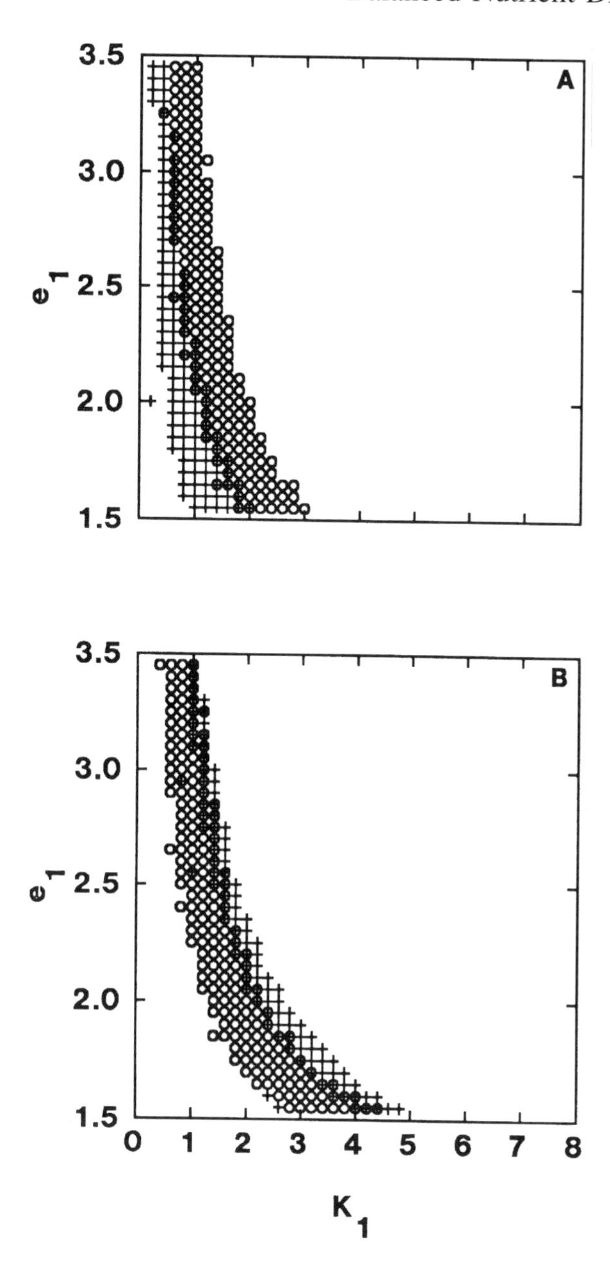

Figure 2.13. Stable combinations of the e_1 versus K_1 parameter space for predator–prey interactions in which the predator either behaves as a generalist (crosses) or facultatively balances its diet (circles). Two scenarios are depicted: (*A*) the poorest prey are profitable enough to allow predator increase, $e_2/h_2 = 1.2$; (*B*) the poorest prey are too unprofitable to permit predator persistence $e_2/h_2 = 0.8$. Other parameter values were as follows: $a = c = d = h_1 = h_2 = r_1 = r_2 = 1$, $z = 4$, $K_2 = K_1$.

that could support a pure generalist predator are unable to support an adaptive predator. The converse is true for simulations in which $e_2/h_2 < d/c$. Accordingly, there was essentially no change in the overall degree of stability of systems with adaptive (balanced diet) versus nonadaptive (no diet balancing) predators.

A view of the zero growth isocline for the system with an adaptive predator shows why there was no stabilizing effect. For most values of N_2, the zero isocline for prey species 1 has a parabolic shape (Fig. 2.14) almost identical to that of a nonadaptive system. When N_2 is particularly low, then the reduction in consumption of prey 1 serves to expand the peak of the parabola upward (implying that it takes more predators to exactly match recruitment by prey). At no point, however, is there any inflection or portion of the isosurface that shows positive acceleration, hence there is no compensatory reduction in prey mortality that could stabilize dynamics. The reduction of predation on prey 1 when prey 2 is rare is therefore a curiosity of little dynamical importance, at least in the context of a system in which there is only indirect interaction between prey species.

Our conclusions are largely consistent with those of Abrams and Shen (1989), who analyzed a similar balanced diet model differing in assuming a linear, rather than type II, functional response and trade-offs between foraging on the two prey types, such as spatial segregation. Abrams and Shen (1989) showed that such a system can be stable, unstable or possess multiple attractors, of which one is always stable.

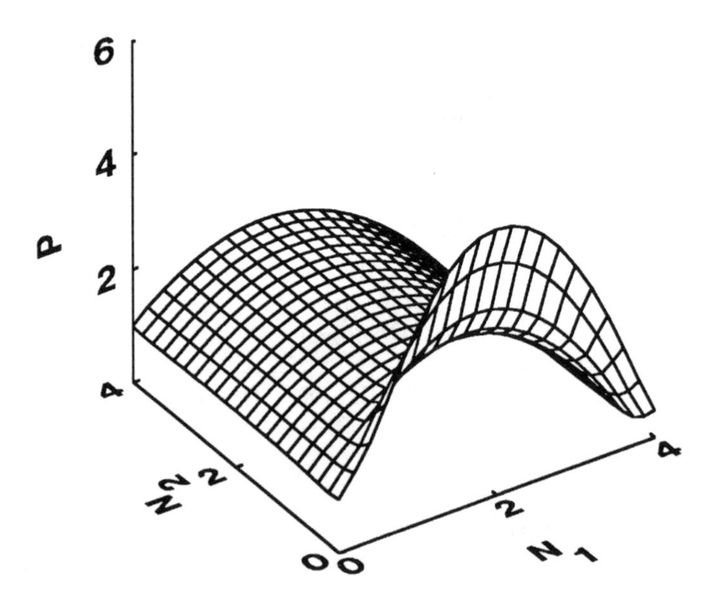

Figure 2.14. Zero growth isocline for a system with a predator facultatively balancing nutrient requirements obtained from two prey with different profitabilities. Parameter values were as follows: $a = c = d = h_1 = h_2 = r_1 = r_2 = 1$, $z = K_2 = K_1 = 4$, $e_1 = 2$, $e_2 = 0.8$.

2.4 SUMMARY

We have only scratched the surface. There are many other diet choice models and predator–prey structures that one could imagine. Hence, any conclusions that emerge must necessarily be tentative and limited to the kind of general systems we have had in mind. Notwithstanding this caveat, several clear tendencies have emerged.

First, adaptive diet selection has at least some potential for exerting a stabilizing influence on predator–prey communities, because the density-dependent forms of choice around critical thresholds imply sudden changes in the probability of attack. In the case of nutrient-maximizing models, the optimal behavioral decision leads to density dependence that can be potentially stabilizing, whereas the inverse density dependence of balanced-diet models is not stabilizing.

Second, whether this potential stabilizing effect is realized depends entirely on the other characteristics, or parameters, of the system in question. In the case of diet choice, the crucial question is whether an ad lib diet of the lowest-ranked prey is sufficient to sustain predators. This is unlikely for most carnivores: It is difficult to imagine a lion that couldn't survive and reproduce on even the lowest ranking item in its savannah environment. It is a more likely proposition for herbivores: Low-ranking items are often of poor enough nutritional quality to preclude survival, let alone reproduction.

Third, stability is enhanced when predators make slightly suboptimal decisions, that is, when partial preferences are pronounced, simply because the sampling errors or other mistakes around the perfect all-or-nothing decision expand the region of stabilizing density dependence. That is, changes in behavior can only exert a stabilizing influence if the range of densities over which behavior is actually changeable is substantial. This condition suggests that the study of the biological sources of partial preferences is of practical importance, if foraging theory is really going to be useful to other branches of community ecology. In this respect, we regard the recent theoretical explorations of the costs of partial preferences and other foraging "mistakes" in state-dependent model structures particularly exciting (McNamara and Houston 1986, 1987; Mangel and Clark 1986; Mangel 1992; Mangel and Roitberg 1992).

Finally, the slight suboptimality implied by partial preference models can lead to surprising changes in the ESS and sometimes allows multiple ESS in the same population. The behavioral literature has long recognized the breadth of alternate behaviors that persist in any population, which is often interpreted in terms of frequency-dependent effects in equilibrium systems, essentially due to individuals playing a game against all other individuals in the population. Our diet selection models suggest that dynamical characteristics could play a vital role in shaping ESS foraging strategies.

3 Prey Defense

The world is full of hostile exploiters trying to take full advantage of their resources. The same selective forces that make consumers efficient at foraging are therefore also responsible for making the resources sufficiently unattractive as food. As the lion becomes more and more efficient at killing zebras, the zebra would become more and more efficient detecting or evading lions. This coevolutionary arms-race analogy is appealing, but nevertheless questionable. Among others, Vermeij (1982), Abrams (1986), and Malcolm (1992) have argued that the details of the mechanisms regulating prey capture (primarily the functional response) and the exact nature of costs and benefits associated with prey defensive actions and predators foraging efficiency must be understood before we can infer an arms-race situation. This does not mean, however, that there is no room for coevolutionary processes in predator–prey interactions. In this chapter, we consider how prey defense may alter the interaction between themselves and their predators. An intriguing aspect of prey defense mechanisms is how sessile organisms, such as plants, may overcome the constant threat of being eaten.

3.1 TYPES OF DEFENSES

We may distinguish between two major modes of defense: avoidance of detection and avoidance of attack once detected. Obviously, they are not mutually exclusive and the former often may involve the latter. The distinction becomes clearer if we compare mobile prey (e.g., most animals) versus immobile prey (e.g., most plants). From a herbivore's point of view, prey are relatively easy to find and attack because plants are generally sessile and apparent. Plants are, however, relatively difficult to handle. Although cropping is not particularly troublesome, nutrient and energy extraction is. It often takes a specialized gut morphology and digestive physiology to make use of plant tissues, at the cost of prolonged handling time in the digestive tract. Carnivores, on the other hand, usually have less trouble processing prey once captured, but prey detection and capture are substantially more difficult. Although this categorization differs somewhat from previous treatments (Endler 1986; Malcolm 1992), it makes the same point: Variation in potential defense strategies is enormous and the exact nature of the behavioral interaction between prey and predator is crucial for understanding the evolution and dynamical implications of antipredator adaptations.

Reduced detection by predators can be achieved in many ways. Prey can indeed be present in the environment but escape by being cryptic. Prey can also be social, which should heighten the risk of detection of the entire group but lessen the risk of individual capture should a predator attack. Prey can also adjust the time they are exposed to predators. Here lies a potential conflict: Minimizing exposure risk necessarily compromises the rate of energy gain. Evasion of predators once encountered imposes other problems. Some animals and many plants are toxic or strongly unpalatable. Physical defensive structures, such as spines or thorns, increase handling time per prey, regardless of whether prey are plants or animals. Finally, prey evasion can render an attempted attack unsuccessful. Such defensive measures have commensurate costs. In some instances, this can be at least partly overcome by having a facultative defense that is induced only when needed. This defense seems to be particularly common among plants, but is also present in certain animal taxa (Karban and Myers 1989; Harvell 1990).

3.2 DEFENSE EFFECTS AND POPULATION PARAMETERS

A number of the foregoing problems are perhaps best illustrated by going back to a simple 2-species model. Recall from Chapters 1 and 2 that the per capita consumption rate of prey can be written

$$X = \frac{aN}{1 + ahN} \tag{3.1}$$

and the corresponding per capita numerical response of the predator population can be expressed as $dP/Pdt = ceX - d$. In the context of predator avoidance or evasion, the parameters a (attack rate or search efficiency), h (handling time per prey), and e (energy content per prey) are of primary interest. The conversion coefficient (c) could just as readily be an appropriate agent for a defensive response, but because it enters in the numerical response in exactly the same way as does energy content, we ignore conversion efficiency here. A decrease in a or an increase in h would each tend to decrease consumption by predators, because $\partial X/\partial a > 0$ and $\partial X/\partial h < 0$. Changes in prey energy content (e) would not directly affect rates of consumption, but might well change the density of predators that can be sustained at equilibrium by a given density of prey. We consider the dynamical implications of these changes in more detail.

Recall from Chapter 1 that no population can grow without limit. This is reflected by the logistic equation specifying density-dependent growth by prey. Following a common convention in life history theory, we can define prey fitness (w) as the per capita rate of prey growth ($[1/N][dN/dt]$). In this particular case, fitness equals per capita reproduction minus the risk of mortality due to predators:

$$w = r(1 - N^*/K) - \frac{aP^*}{1 + ahN^*} \tag{3.2}$$

In order to maximize prey fitness, the rightmost term of equation 3.2 must be minimized. From Chapter 1, we already know the quantities N^* and P^*. Substituting these equilibria into equation 3.2 and rearranging yields the total prey mortality rate at equilibrium $Y^* = XP^*/N^*$:

$$Y^* = r(1 - \frac{d}{Ka(ce - dh)}) \tag{3.3}$$

Equation 3.3 is of course identical to $r(1 - N^*/K)$, which follows from the evaluation at equilibrium. We can use equation 3.3 to examine the relative effects of adjusting the available parameters. That is, we are interested in comparing the following partial derivatives:

$$\frac{\partial Y^*}{\partial a} = \frac{rd}{Ka^2(ce - dh)} \tag{3.4}$$

$$\frac{\partial Y^*}{\partial e} = \frac{rcd}{Ka(ce - dh)^2} \tag{3.5}$$

$$\frac{\partial Y^*}{\partial h} = \frac{rd^2}{Ka(ce - dh)^2} \tag{3.6}$$

Recall that the quantity $(ce - dh)$ must be positive in order to have a positive N^*, and that we are only interested in the magnitude (absolute value) of the partials, not their sign. Given those restrictions, we may draw the following conclusions: for a given value of a, $\partial Y^*/\partial e > \partial Y^*/\partial h$ if $c > d$; for a given e, $\partial Y^*/\partial h > \partial Y^*/\partial a$ if $N^* > 1/a^2$; for a given h, $\partial Y^*/\partial e > \partial Y^*/\partial a$ if $N^* > d(ca^2)$. Translated into words, the first case says that all other things being equal, decreasing energy content reduces prey mortality more than an increase in handling time, provided that $c > d$. By the same token, increasing handling time reduces prey mortality more than decreasing search efficiency, provided that prey are particularly common at equilibrium. Finally, if handling time cannot be adjusted, decreasing energy content reduces prey more than to decrease search efficiency, provided once again that prey are particularly common.

Many plant strategies seem to fit these general predictions, although rigorous experimental tests are hard to imagine. The relatively high attack rate on plants would generally call for adjustments to energy content and handling time rather than further changes in search efficiency (Crawley 1983; Lundberg and Palo 1994). This also reiterates Abrams's (1986) conclusion that unlimited arms-races would probably not come into effect because the defensive actions taken by prey will depend on both the previous suite of antipredator strategies and environmental circumstances.

3.3 PARAMETER EFFECTS ON DYNAMICS

The above arguments only apply under equilibrium conditions, i.e., when N^* and P^* exist and population densities are static. This is obviously questionable, because attack rate, energy content, and handling time also influence dynamical properties of the community. If we retain the same predator–prey system, then we know that $N^* = d/(a(ce - dh))$ and $P^* = (r/a)(1 - N^*/K)(1 + ahN^*)$. If search efficiency or energy content were decreased by prey, or h increased commensurately, the equilibrium density of prey would increase. The effect of defensive actions taken by prey on P^* is more complicated. Figure 3.1 illustrates how changes in a, e, and h affect the equilibrium density of predators. Figure 3.2 shows how changes in parameters affect the joint equilibrium of prey and predators. The local stability of this predator–prey system has the nice property of being fully determined by the slope of the prey isocline at the equilibrium point (N^*, P^*) (cf. Chapter 1). If we call this slope f, then

$$f = \left.\frac{\partial P^*}{\partial N}\right|_{N^*,P^*} = \frac{r(ahK(ce - dh) - ce - dh)}{aK(ce - dh)} \tag{3.7}$$

As attack rate and energy content decrease, so does f, i.e., the equilibrium becomes more stable. We get the same result if handling time increases. Hence, all other things being equal, the same natural selection processes favoring reduced search efficiency, reduced energy content, or increased handling time should have a ubiquitously stabilizing effect on trophic dynamics.

The foregoing reasoning leaves no room for behavioral decision making, pointing only to general effects of changes in demographic parameters. Decisions are only meaningful, of course, if there are options and trade-offs, and defensive strategies are no exception. This topic has been widely debated in the literature, especially with respect to plants (Fagerström et al. 1987; D. G. Brown 1988; Herms and Mattson 1992; Simms 1992; Clark and Harvell 1992; Adler and Karban 1994; Åström and Lundberg 1994). Similar lines of research have examined time allocation by animals as an antipredator strategy (Abrams 1984, 1989, 1990a, 1994a; Abrams and Matsuda 1993; J. S. Brown 1988, 1992; Mitchell and Brown 1990; Werner 1992; Matsuda et al. 1993; Werner and Anholt 1993). In the following, we examine the nature of such trade-offs in more detail. Our point of departure is again the familiar two-species predator–prey system. We then extend the analysis by including another trophic level (appropriate for example for a carnivore–herbivore–plant system), which alters the cost and benefit functions.

3.4 TIME ALLOCATION

It is perhaps debatable whether we should call time allocation a defense in the strict sense, but it is nevertheless a tactic that could potentially function in a

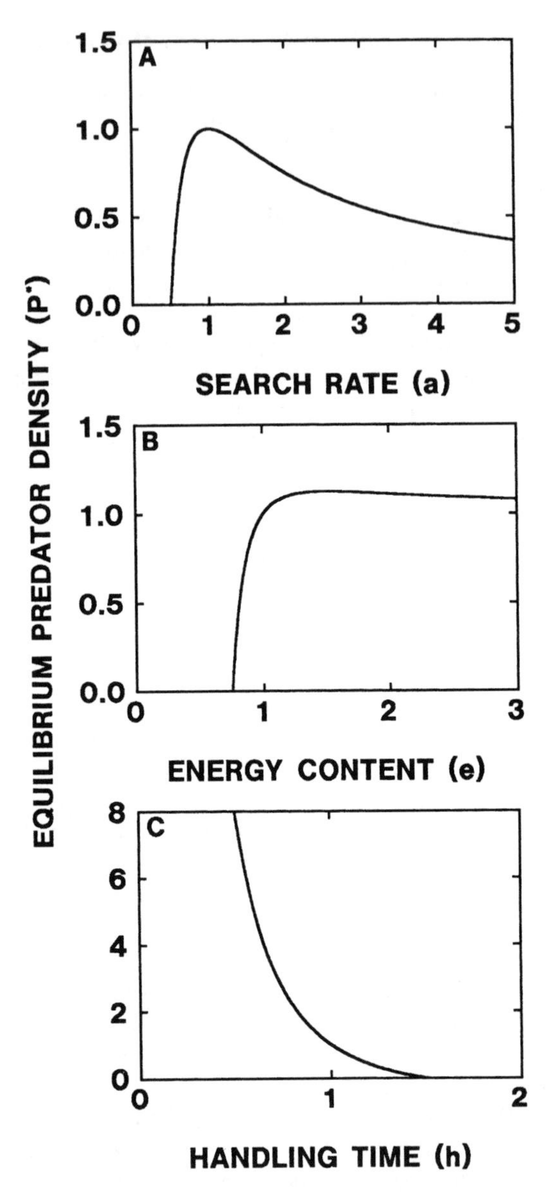

Figure 3.1. The equilibrium density of predators in the predator–prey system described by equations 1.1 and 1.2 in relation to attack rate (a), prey energy content (e), and handling time of prey (h). Other parameter values: a = r = h = c = e = 1, d = 0.5, K = 2.

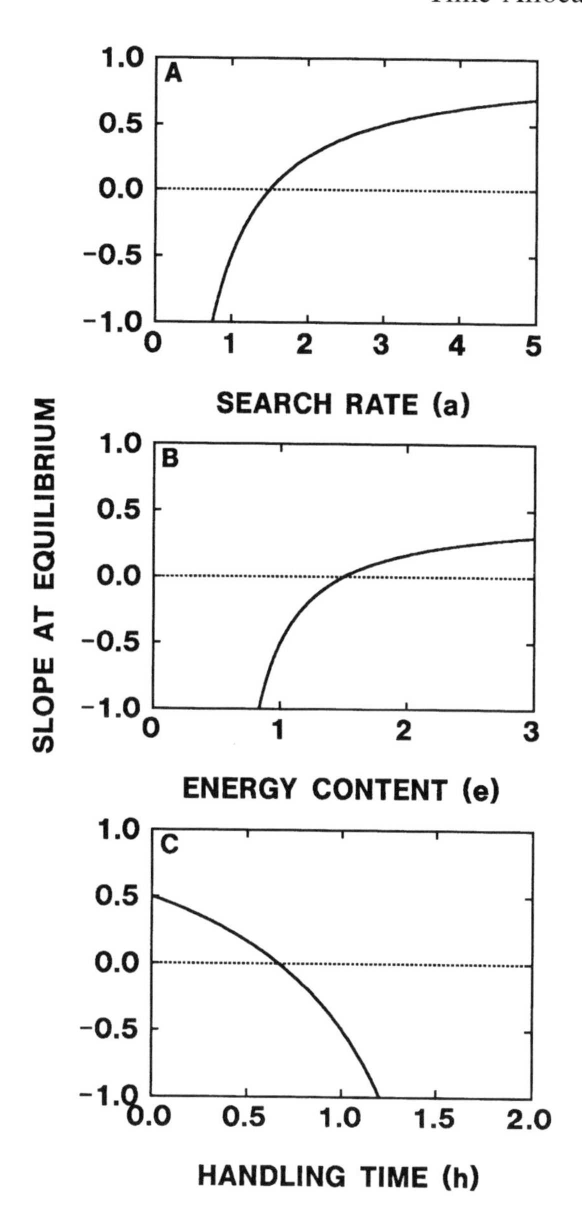

Figure 3.2. The change of the slope of the prey isocline at the equilibrium point in the phase plane. If the slope is < 0, the equilibrium in this system is locally stable. Same parameter values as in Fig. 3.1.

defensive context (Curio 1976; Waldbauer et al. 1977; Kidd and Mayer 1983; Waldbauer 1988; Malcolm 1992). In addition to a vast number of theoretical treatments of the problem (e.g., Milinski and Heller 1978; Abrams 1982, 1983, 1986, 1990a; Sih 1984; Lima 1985; Oksanen and Lundberg 1995), there is also substantial empirical evidence for the trade-off between energy gain from foraging and the risk of predation during this occupation, including insects, fish, birds, and mammals (e.g., Stein and Magnuson 1976; Sih 1980, 1982; Werner and Mittlebach 1981; Werner et al. 1983; Lima 1985; Lima et al. 1985; J. S. Brown 1988; Lima and Dill 1990; Werner 1992; Korpimäki et al. 1994). The behavioral response to predation risk varies considerably across systems. Predation risk may cause adjustments in foraging speed (Werner and Anholt 1993), instantaneous intake rate (Sih 1982; Lima 1985), diet choice (Werner and Mittlebach 1981; Werner et al. 1983a; Brown and Mitchell 1989), patch use (Werner and Mittlebach 1981; J. S. Brown 1988; Brown et al. 1992a,b) and reproductive decisions (Korpimäki et al. 1994). The latter example is particularly illustrative of the kind of tradeoffs we are envisioning. Korpimäki et al. (1994) studied the reproductive investment of voles (*Clethrionomys* and *Microtus*) under predation risk from mustelids. Litter size of bank voles was negatively correlated with densities of both stoats and weasels (Fig. 3.3). The results were supported by both observational and experimental data, i.e., litter size responded not only to natural variation in mustelid density, but also to experimental removal of predators. Korpimäki et al. (1994) concluded that since the vole antipredator behavior also occurred on a large spatial scale, as evident from the observational data, such behavioral adjustments resulting in significant population parameter changes could be crucial to the dynamics of the vole–mustelid system.

We now examine the dynamical effects of such antipredator behavior in more detail. We commence by analyzing a 2-species predator–prey system in which the prey must trade off foraging gains against the risk of being preyed upon while foraging. This system is then extended to a 3-species food chain in which the predator itself is subject to predation risk.

3.4.1 Risk-Sensitive Prey

For simplicity, we let prey benefit indirectly from foraging, by making the maximum per capita growth rate a function of the proportion of total time devoted to foraging (τ). If $\tau = 0$, then the animal allocates no time at all to foraging, hence not acquiring any resource for reproduction. If $\tau = 1$, then foraging is maximized, as is the per capita growth rate. Ignoring the fact that the exact shape of the cost and benefit functions can have a profound effect on optimal behavioral solutions (Abrams 1982, 1991; Åström and Lundberg 1994), we let r and τ be linearly related, i.e.,

$$r(\tau) = \tau r_{max} \qquad (3.8)$$

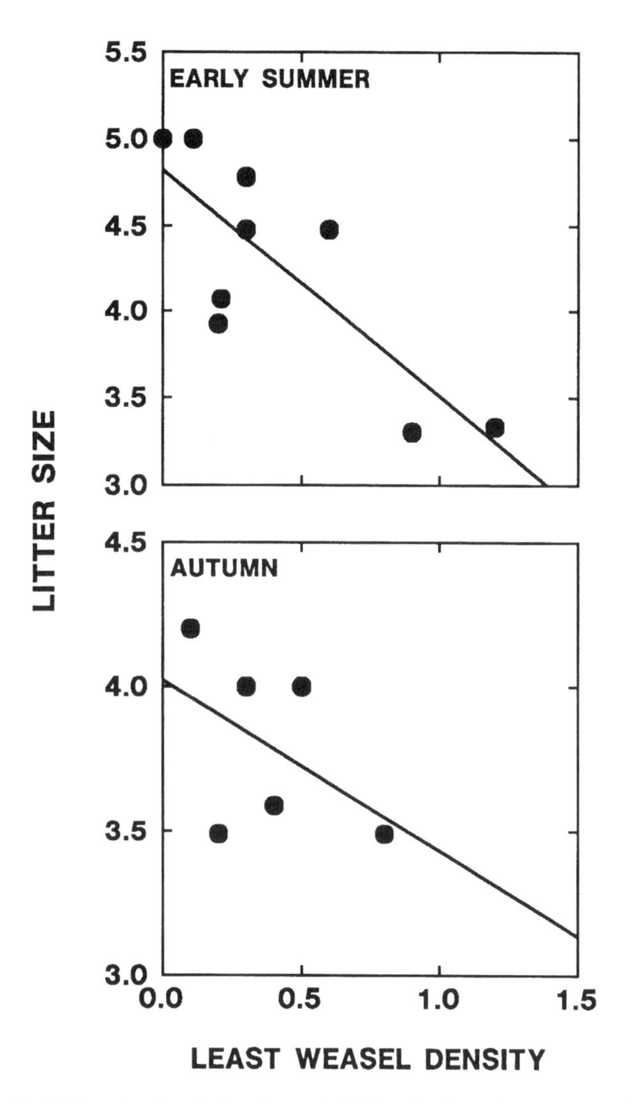

Figure 3.3. Vole litter size in relation to mustelid density in early summer and in the fall in Finland. Increasing perceived predator density suppresses breeding in female bank voles (*Clethrionomys glareolus*). From data of Korpimäki et al. 1994.

The risk of predation is also a function of τ. Assume that active foraging increases the frequency with which predators and prey come into contact, according to the same mass action principles that underlie the functional response. The rate of effective search (a) would increase with τ, in the simplest case according to

$$a(\tau) = a_{min} + \tau(a_{max} - a_{min}) \tag{3.9}$$

where $a_{\min}$ (>0) is the minimum rate of search (due to only predator movements) and $a_{\max}$ = the maximum rate of search, when both predators and prey are continually moving. We now make the bold assumption that all the individuals in the prey population adopt the same τ. Hence, the full dynamical system becomes

$$\frac{dN}{dt} = r(\tau)N(1 - N/K) - \frac{a(\tau)NP}{1 + a(\tau)hN} \tag{3.10}$$

$$\frac{dP}{dt} = \frac{cea(\tau)NP}{1 + a(\tau)hN} - dP \tag{3.11}$$

Maximizing foraging activity maximizes food acquisition and potential growth, but also maximizes the risk of predation. Note that we are here assuming that the predator does not adjust its foraging in response to the behavioral strategy played by prey individuals. Minimizing τ, on the other hand, leads to inevitable death. Let us define prey fitness, w, to be the per capita rate of growth $(1/N)(dN/dt)$. We are interested in knowing whether there is a τ that maximizes fitness. It turns out that $w(\tau)$ has no local maximum, but potentially a minimum (which can be readily verified by noting that the second derivative of w with respect to τ is greater than 0). This means that either $\tau = 1$ or $\tau = 0$ is maximizing fitness. The two alternative fitness maximizing solutions are:

$$\tau = 1 \text{ if } r_{max}(1 - N/K) > \frac{a_{max}P}{1 + a_{max}hN} - \frac{a_{min}P}{1 + a_{min}hN} \tag{3.12}$$

$$\tau = 0 \text{ if } r_{max}(1 - N/K) < \frac{a_{max}P}{1 + a_{max}hN} - \frac{a_{min}P}{1 + a_{min}hN} \tag{3.13}$$

Clearly, the optimal behavioral solution will jump between 0 and 1 depending on population densities of predators and prey. We can express this switch point by making one population density a function of the other, i.e., let ρ be the predator density at which the optimal τ switches from 0 to 1, after rearrangement yielding

$$\rho = \frac{r(1 - N/K)}{\dfrac{a_{max}}{1 + a_{max}hN} - \dfrac{a_{min}}{1 + a_{min}hN}} \tag{3.14}$$

If $P > \rho$ then $\tau = 0$ and from equation 3.13 we see that dN/dt is always < 0 and $dP/dt < 0$ if $N < d/(a_{\min}(c - dh))$ whereas $dP/dt > 0$ if $N > d/(a_{\min}(c - dh))$. On the other hand, if $P < \rho$, then $\tau = 1$ and equation 3.12 specifies the usual rates of change for both predators and prey. This situation leads to a rather strange set of zero growth isoclines (see Chapter 1 for an introduction) (Fig. 3.4). If predators are sufficiently common and prey density happens to be above N^*,

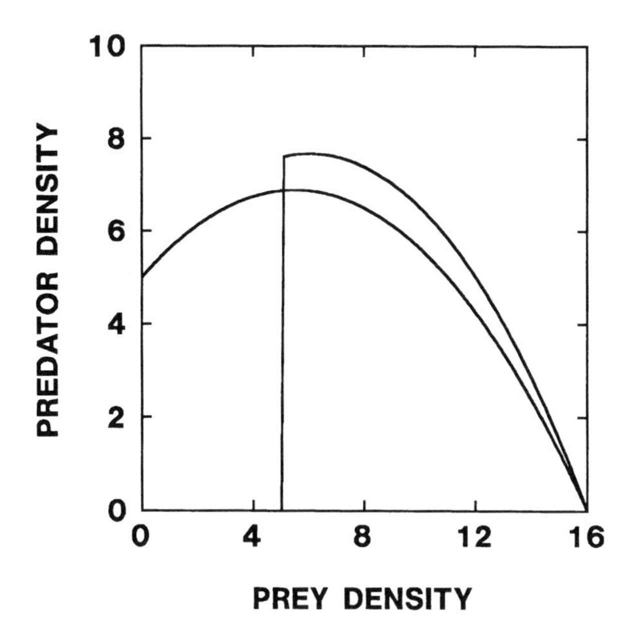

Figure 3.4. The phase plane for predators and prey specified by equations 3.10 and 3.11. The predator isocline is bent downward at high predator densities because when predators are common, prey cease foraging and both predators and prey populations decrease. The facultative adjustment of the foraging time τ thus is stabilizing this system. Parameter values as in Fig. 3.1.

prey cease foraging and both prey and predator populations immediately begin to decrease. This has the same effect as interference among predators (Chapter 6). Even if prey are abundant enough for the predator population to continue to grow, the adaptive adjustment of foraging time sharply reduces the attack rate to a_{min}. This happens not only when predator density is high, but also when prey density is high, making the per capita risk of prey mortality sufficiently large to switch τ from 1 to 0. Thus, the predator isocline is bent downward toward K for prey densities larger than N^*.

Just as the 0–1 rule for diet choice is unrealistic for several reasons (Chapter 2), so is the 0–1 rule of defensive time allocation. We modified the discontinuous switch of τ such that

$$\tau = \frac{exp(z\rho)}{exp(z\rho) + exp(zP)} \tag{3.15}$$

with z a parameter determining the degree of conformance between the sigmoid continuous function and the idealized step function (see Chapter 2 for a more thorough discussion of this function). If $P = \rho$, then $\tau = 0.5$. As P increases above ρ, τ tends to zero and as P decreases below ρ, τ tends to unity.

Not surprisingly, a system with adaptive foraging time is notably more stable than a null system, in which $\tau = 1$ for all N and P. Figure 3.5*A* and 3.5*B* show examples of time series of systems with and without adaptive foraging time, and in Fig. 3.6 variability of the systems is assessed over a wide range of parameter values. As in most of our examples of facultative behavioral adjustments, the stabilizing effect derives from the fact that there is a threshold prey or predator density in the neighborhood of which there is positive density dependence in prey mortality rates. The breadth of this region of density dependence stems from the degree of conformity between the realized behavioral switch and the idealized step pattern.

The situation of coexisting adaptive and nonadaptive prey populations can be examined in similar fashion to that of predators faced with adaptive diet selection (Chapter 2). Consider two prey populations that share a common predator that feeds on each prey in direct proportion to prey relative abundance. One prey species shows no facultative defensive response to predation risk, whereas the other prey species does respond. Since the two prey populations do not differ in any respect other than defensive behavior, there is no need for diet choice by the predator. If the joint equilibrium is stable, then coexistence is possible as long as equation 3.15 leads to $\tau > 0$, with the relative abundance of the two populations proportionate to τ. If the equilibrium is not stable, however, the populations fluctuate in a manner reminiscent of the coexisting predator populations in Chapter 2 (Fig. 3.5C), for exactly the same reasons.

3.4.2 Three-Link System

We now extend the previous 2-link system to a 3-link system: a carnivore that feeds on a herbivore that feeds in turn on plants at the lowest trophic level. The herbivore must balance the risk of being attacked by the carnivore against its own rate of energy gain. This formulation allows us to be a little more explicit about how costs and benefits are related. Let τ be the fraction of the herbivore's time devoted to foraging. If the functional responses of both the herbivore and the carnivore are monotonically decelerating, then the full three-link system can be expressed as

$$\frac{dN}{dt} = rN(1 - N/K) - \frac{a(\tau)NP}{1 + a(\tau)hN} \tag{3.16}$$

$$\frac{dP}{dt} = \frac{ea(\tau)NP}{1 + a(\tau)hN} - dP - \frac{A(\tau)PC}{1 + A(\tau)HP} \tag{3.17}$$

$$\frac{dC}{dt} = \frac{EA(\tau)PC}{1 + A(\tau)HP} - DC \tag{3.18}$$

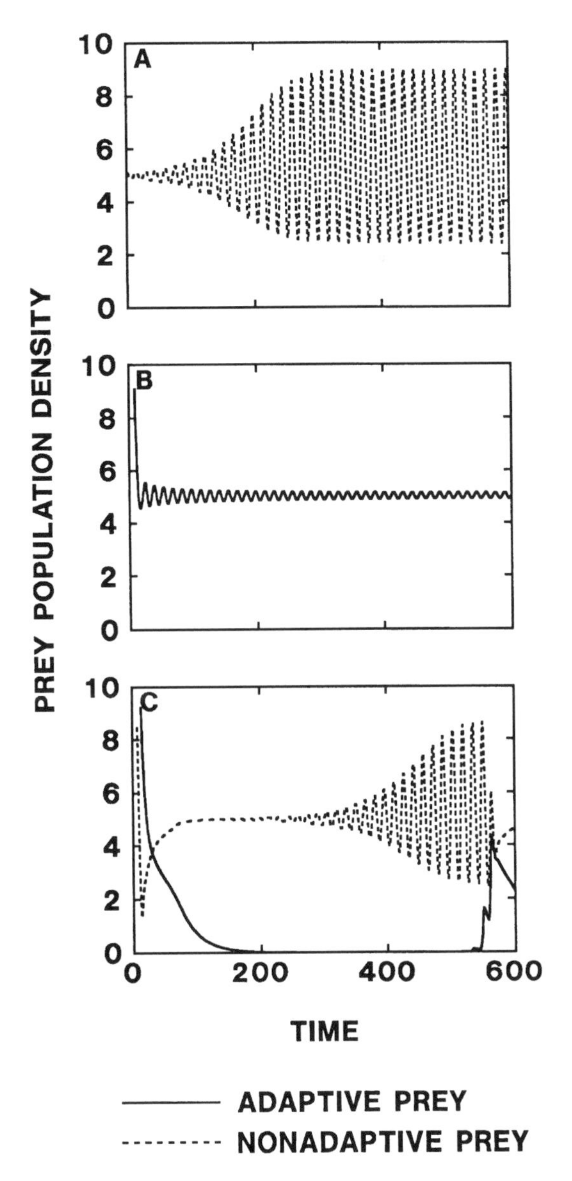

Figure 3.5. The time series of the prey population in a predator–prey system specified by equations 3.10 and 3.11 for nonfacultative (*a*) and facultative (*b*) time allocation. In (*c*), the two strategies are coexisting with a shared predator. Parameter values were $r = h = c = 1.0$, $d = 0.5$, $a_{max} = 0.2$, $K = 16$.

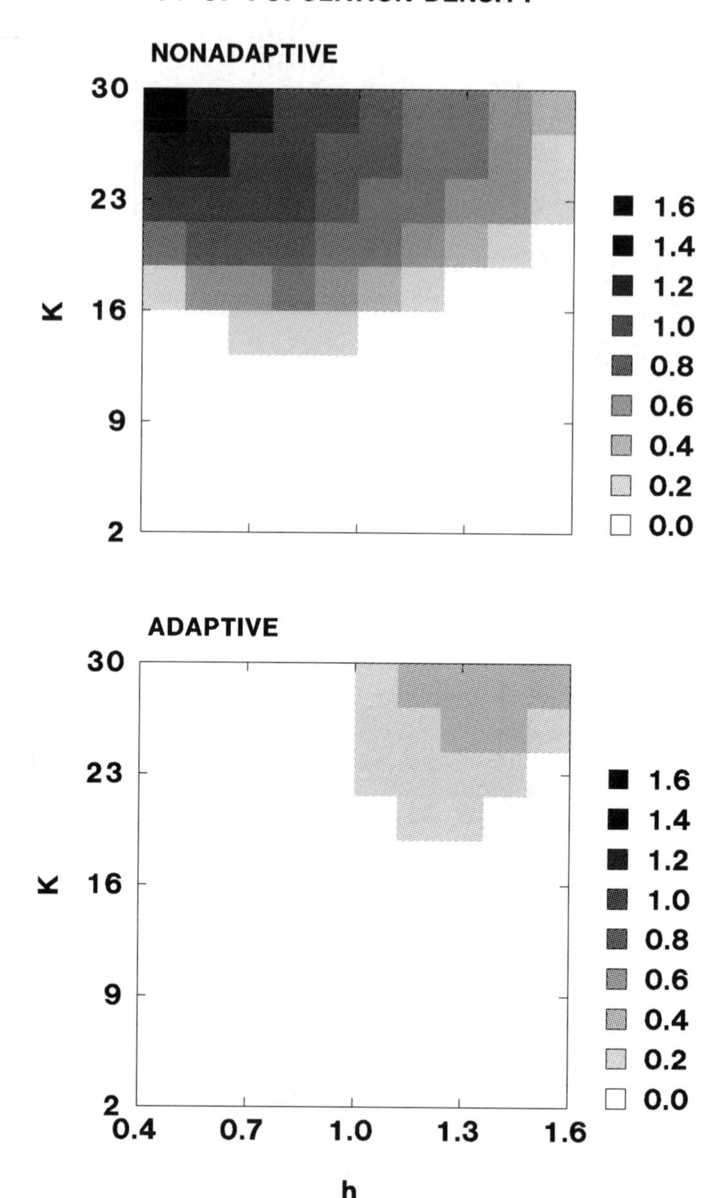

Figure 3.6. Prey population variability (coefficient of variation) in relation to handling time (h) and carrying capacity (K) for nonfacultative and facultative time allocation in the prey population. The CV of population density is notably lower when foraging time is facultatively changed according to prey and predator density for most parameter combinations. Other parameter combinations as in Figure 3.5.

where A = the area searched per unit time by the carnivore, H = the handling time of each herbivore captured by the carnivore, E = the energy content of each herbivore, and C = carnivore population density. Note that search rates by both herbivores and carnivores (a and A) are functions of τ, such that $A = a_{\min} + (a_{\max} - a_{\min})\tau$, which is analogous to the 2-link case above. By definition, τ only takes values between 0 and 1, therefore $a_{\min} \leq A \leq a_{\max}$. The optimization problem for the herbivore is thus analogous to the 2-link case, but here the decision of optimal τ-value for the herbivore also affects both plants and carnivores. Let herbivore fitness, w, be defined by $(1/P)(dP/dt)$. It turns out that $w(\tau)$ has no maximum and that the optimal decision will be either 0 or 1, as in the two-dimensional system. The switch between $\tau = 0$ and 1 happens when

$$C = \frac{ceaN}{(1 + ahN)\left(\frac{a_{max}}{1 + a_{max}HN} - \frac{a_{min}}{1 + a_{min}HN}\right)} \tag{3.19}$$

The stabilizing effect of facultative time allocation is considerably less pronounced in a 3-link system than in a 2-link system. Figure 3.7 summarizes the stability properties of the system with and without facultative changes in foraging time allocation. Variation in herbivore density is only marginally less in the adaptive system than in the null model, but variation in the other trophic levels is less pronounced in the facultative system than in the nonfacultative system. Hence, the three trophic levels respond differently, as shown in Fig. 3.8. Temporal variation in population densities of plants, herbivores, and carnivores are plotted against carrying capacity (K). As K increases, so does variability of each population, as we leave the region of stable equilibria and we get stable limit cycles. For the herbivore population, facultative decision making hardly affects population variability (Fig. 3.8*B*). Plants have the strongest response (Fig. 3.8*C*), in which facultative changes in foraging behavior by the herbivores dampens fluctuations by plants at high carrying capacities. A weaker response is found in the carnivore population (Fig. 3.8*A*). So even though facultative time allocation doesn't influence variability in the actor, it can have more substantial effects on other players.

Another notable effect on food chain dynamics is the response of mean population density as the equilibrium conditions of the system change. For example, as we enrich the environment by increasing prey carrying capacity (K), the populations at the different trophic levels respond very differently. This has been noted in several food chain models (Abrams 1993; Abrams and Roth 1994*a, b*; Lundberg and Fryxell 1995) and this model is no exception. In the null model, plant and herbivore populations respond to increased K by increasing mean densities, whereas the carnivore density decreases as K gets sufficiently high (Fig. 3.9) (Abrams and Roth 1994*b*). In the model with facultative adjustment in herbivore foraging time, all three trophic levels respond positively to enrichment. These differences could greatly alter our interpretation of food chain responses to environmental changes. It also emphasizes the potential higher order effects

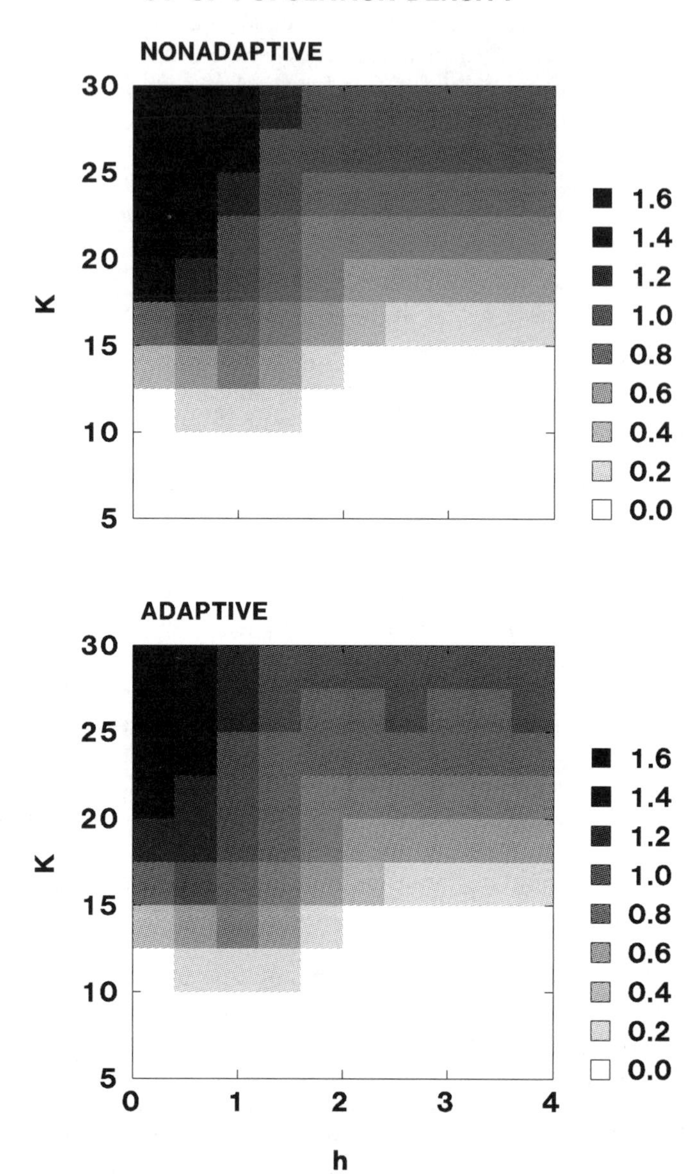

Figure 3.7. Herbivore population variability in relation to handling time (h) and carrying capacity (K) for nonfacultative and facultative time allocation when the focal species (the herbivore) is embedded in a 3-link food chain (eqns. 3.16–3.18). In this situation, the stabilizing effect of facultative time allocation is small (cf. Fig. 3.5). Parameter values were $r = e = 1.0$, $a_{max} = 0.2$, $d_1 = 0.5$ (herbivore density independent death rate), $D_2 = 0.3$ (carnivore density independent death rate).

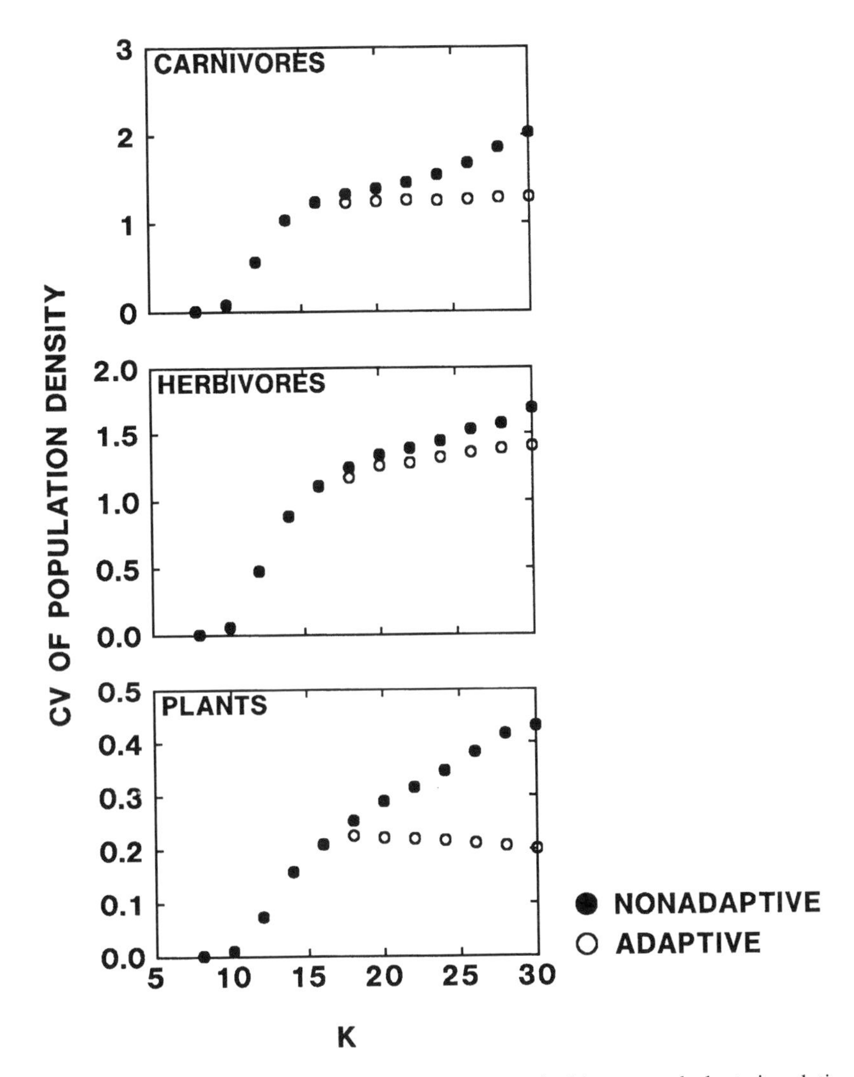

Figure 3.8. The population variability for carnivores, herbivores, and plants in relation to plant carrying capacity (K) for nonfacultative (●) and facultative (○) time allocation. Parameter values as in Figure 3.7.

that behavioral decisions may cause and that behavioral mechanisms should always be considered when trying to interpret not only the dynamics of single or few populations, but also larger food chains or food webs (Abrams 1993; Lundberg and Fryxell 1995).

The differential effect of facultative adjustment in herbivore foraging time on herbivore and carnivore intake rates is illustrated in Figure 3.10. Here, the intake

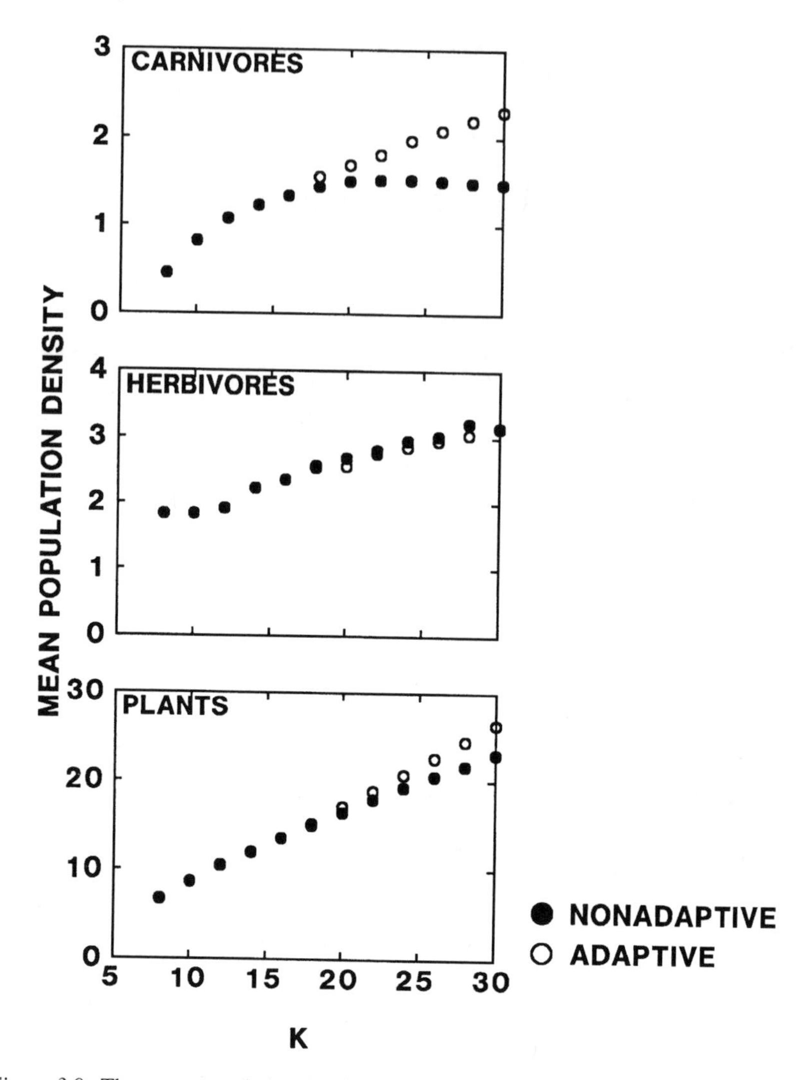

Figure 3.9. The mean population density for carnivores, herbivores, and plants in relation to plant carrying capacity (K) for nonfacultative (●) and facultative (○) time allocation. Parameter values as in Figure 3.7.

rate per individual consumer (i.e., the functional response) is shown in relation to both plant and herbivore densities. The herbivore has a sigmoid functional response to plant density, which follows directly from the fact that herbivores switch from foraging to nonforaging when plants are rare. The feeding rate by herbivores is also affected by herbivore density, such that increased herbivore density actually facilitates herbivore feeding. This ameliorative effect is due to

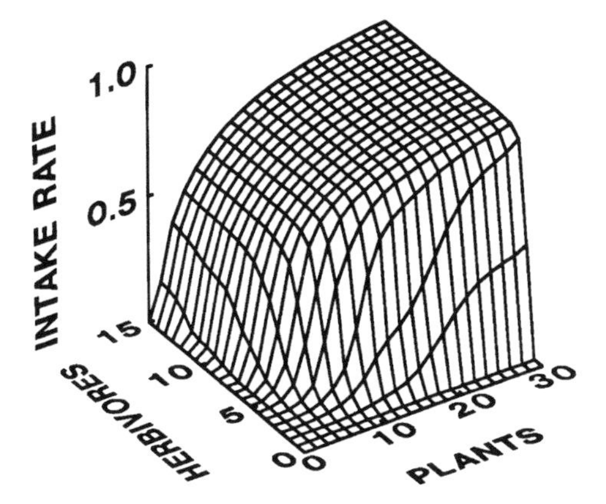

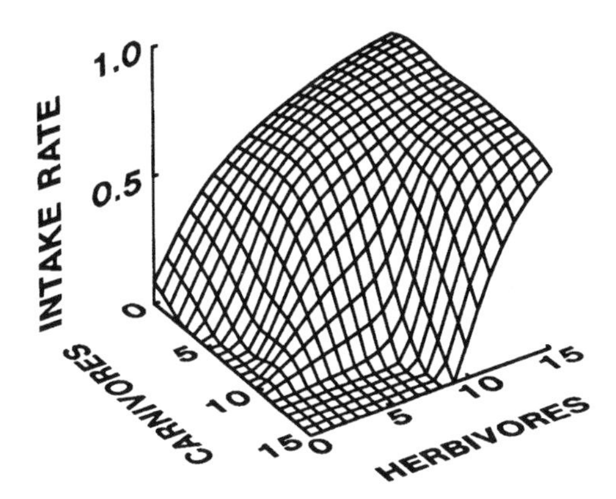

Figure 3.10. The per capita intake rate for herbivores and carnivores in relation to plant and herbivore density, and herbivore and carnivore density, respectively. For herbivores, the per capita intake rate increases with both plant and herbivore density. For carnivores, intake rate increases with herbivore density, but decreases with the density of itself. Parameter values as in Figure 3.7.

the fact that as herbivore density increases, reduction in risk makes feeding a more attractive proposition. The carnivore has a sigmoid functional response to herbivore density, because of herbivore adjustments of τ. On the other hand, increasing carnivore density induces herbivores to cease foraging, thereby reducing carnivore attack rates.

Mitchell and Brown (1990) conducted a foraging experiment with desert rodents (*Dipodomys merriami*) provided a mixture of seeds and sand on experimental feeding trays. Their results clearly showed a sigmoid functional response of the type one might expect on the basis of optimal time allocation. Mitchell and Brown argued that this response was due to the high metabolic costs of foraging outweighing the cost of inactivity at low seed densities, but predation risk could also act as the fitness cost that reduces the time allocated to foraging (J. S. Brown 1988).

The model systems we have outlined share many features with a host of similar models (e.g., Abrams 1984, 1990*a*, 1991, 1992; McCauley et al. 1988, 1993; Gatto 1991). The conclusion that emerges, and that is worth emphasizing, is that adaptive responses to the trade-off between predation risk and energy gain may significantly alter the way we both model and interpret trophic communities. We find it encouraging that theory and data are both well developed in this field and that time allocation has been well demonstrated. The full suite of dynamical implications in real world systems is, however, yet to be explored.

3.5 OPTIMAL DEFENSE

Although the previous treatment focused on foraging time allocation as a form of anti-predator defense, the structure of the models is sufficiently general to encompass other similar fitness trade-offs. We now turn to a somewhat different scenario: prey defense. We make no explicit assumption about exactly how the defense investment affects prey fitness, only that it does have a negative impact on prey growth rate. Moreover, we only assume that the defense decreases the predator's rate of consumption, much in the same way as in the previous section. We outlined some of the potential effects on different population parameters at the beginning of this chapter. We use a model first suggested by Ives and Dobson (1987). They did not specify the exact biology of the system they analyzed and we too make use of that generality.

Our basic assumption is again that defense reduces the per capita risk of mortality through predation at a price of diminished rate of prey growth or increased mortality due to factors other than predation. In the case of time allocation, the cost of the defensive action can and has been measured. For instance, Werner (1992) and Korpimaki et al. (1994) have shown that animals pay a direct fitness cost (growth and/or reproductive output) by reducing predation risk. J. S. Brown (1988) has also developed models and techniques to measure the missed opportunity cost when the animal decides to defend itself from (or rather avoid) predation. Defense costs have been notoriously difficult to measure in plants, although the benefits are well documented. Much of plant defense theory rests on the assumption of a physiological cost of plant defense, manifested by reductions in growth or reproduction (Fagerström et al. 1987; Skogsmyr and Fagerström 1992; Herms and Mattson 1992). The evidence, however, is equivocal. In a recent review, Simms (1992) concluded that costs can be significant but that the circumstances under which antiherbivore defenses also incur fitness costs vary considerably. Although plants are sessile, the risk of herbivory may vary greatly, not least due to changing herbivore densities. It has been suggested that under such variable predation risk, induced defense strategies would be favored (Clark and Harvell 1992; Riessen 1992; Adler and Karban 1994; Åström and Lundberg 1994).

Many invertebrates also seem to use the same strategy (Harvell 1990; Clark and Harvell 1992). For example, freshwater cladocerans such as *Daphnia* form spines when exposed to potential predators. Walls and Ketola (1989) showed that predator presence indeed is a cue for spine production and that spine production is associated with fitness costs. They subjected *D. pulex* to one of its main predators (*Chaoborus* larvae, a dipteran) and observed that induced spine production both postponed first reproduction (from fifth to sixth instar) and slowed growth. Slower growth of the spined individuals leads to smaller individuals in a given instar, leading to reduced clutch size. There is ample evidence that spines do function as a defense against predation. Figure 3.11 shows the results from an experiment with *D. pulex* preyed upon by *Chaoborus americanus*. Spine production had no effect for the smaller instars that were easily captured anyway, but was very efficient in larger instars (Havel and Dodson 1984). An experiment with bryozoans (Harvell 1990) has also shown that the induction of defense is not necessarily

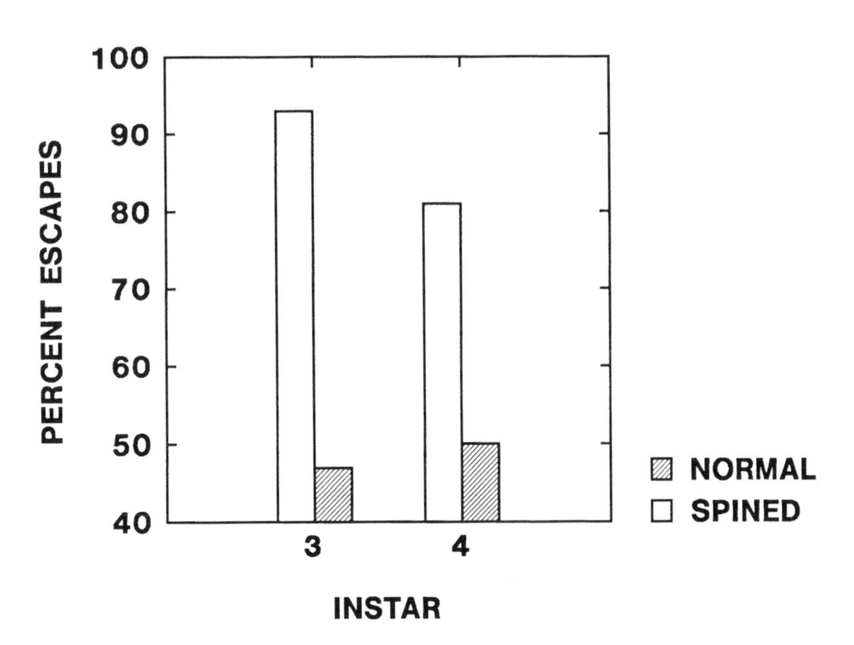

Figure 3.11. The proportion of unsuccessful attacks by *Chaoborus* on spined and normal morphs of *Daphnia pulex*. Clearly, the spined morph escapes predation more efficiently than the normal one. From data of Walls and Ketola (1989).

an all-or-none-process (Fig. 3.12). Bryozoan colonies were exposed to different levels of extracts from a major predator (nudibranchs). The level of defense (spine production, known to reduce nudibranch foraging success) increased continuously with predator density (in this experiment predator extracts). Thus, one beauty of induced defense is that the exact level of defense can be adjusted to prevailing risk.

Theoretical studies of induced plant defenses in relation to both mean and variance of risk of herbivory (Adler and Karban 1994; Åström and Lundberg 1994) suggest that there may be no obvious advantage of induced defenses unless certain conditions are met. Among the most important of these are the mean risk of herbivory, how variable this risk is over time, and the magnitude of defense costs. Certain life history traits, such as the ratio of juvenile to adult survival rate and semelparity versus iteroparity also seem to play important roles. These conditions may also explain why both the costs and benefits of defensive tactics are often ambiguous in invertebrates (Adler and Harvell 1990).

Few studies have attempted to link induced strategies with dynamic population models (Hakala and Haukioja 1975; Fox 1981; Edelstein-Keshet and Rausher 1989; Lundberg et al. 1994). In plants, not all putative induced responses have been shown to be effective as deterrents (Karban and Myers 1989). Although time scales may vary, induced plant defenses commonly lag behind the triggering

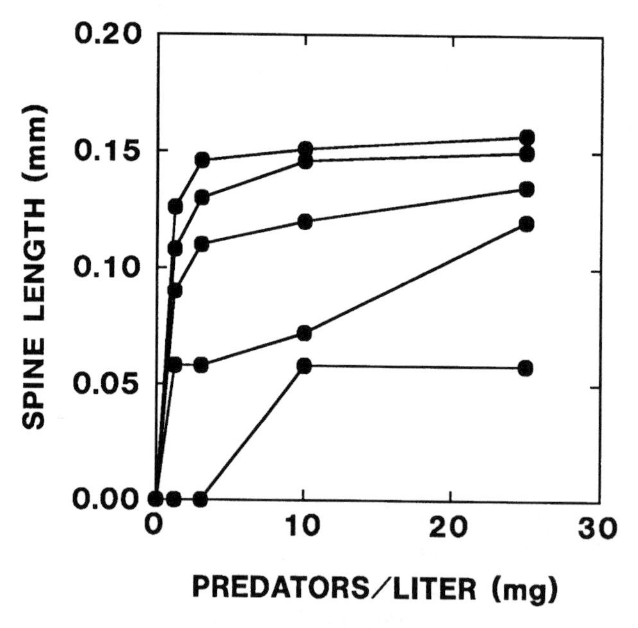

Figure 3.12. The induction of spines (spine length) in relation to perceived risk of predation (the concentration of nudibranch extracts) in the bryozoan *Membranipora membranacea.* The five different curves relate to different bryozoan colonies (genotypes). Data from Harvell (1990).

mechanism, which is probably why most defensive induction models and available data (Crawley 1983; Karban and Myers 1989; Lundberg et al. 1994) indicate that induced defenses tend to be dynamically destabilizing. Induced defenses in most plants and in even some bryozoans (Harvell 1990) occur after the first, nonlethal attack. In most animals, on the other hand, perceived predation risk is enough to induce defense. For example, the freshwater rotifer *Keratella testudo* exposed to water filtrates of cultures of its normal predators (e.g., the cladoceran *Daphnia pulex,* also a *Keratella* competitor) induced posterior spines known to be protective (Stemberger and Gilbert 1987). Water filtrates from herbivorous rotifers and dipteran larvae did not, however, induce spine formation (Stemberger and Gilbert 1987).

3.5.1 Perfectly Timed Induced Defense

Let δ be a nonnegative parameter representing the defense level. When $\delta = 0$, then no resources at all are invested in defense and no price is paid. If $\delta > 0$, then there is a benefit in terms of reduced per capita predation risk that is related to δ, and a cost of reduced per capita reproduction rate that is proportional to δ. Following Ives and Dobson (1987), we assume that the benefit is nonlinearly related to the level of defense investment, but that the cost is directly proportional to δ. Unlike the previous systems we have been looking at in this chapter, nonlinear functions allow for the possibility of an optimal level of defense, δ^*, at which prey fitness is maximized. The full system can then be written

$$\frac{dN}{dt} = rN\left(1 - \frac{N}{K}\right) - \delta N - \frac{qe^{-\theta\delta}NP}{1 + sN} \tag{3.20}$$

$$\frac{dP}{dt} = \frac{cqe^{-\theta\delta}NP}{1 + sN} - dP \tag{3.21}$$

We attempt to keep this treatment as general as possible, therefore let N denote prey or plant population density and P predator or herbivore density. In keeping with the original model, we have a slightly different formulation of the functional response in eq. 3.20 from the earlier ones. Here, q scales the predation rate and s sets the rate at which predators become satiated (which means that in eqn. 3.20 s/q is equal to h in previous ones). The parameter θ reflects the efficiency of the defense, with large θ indicating efficient defense.

Let us again define fitness, w, as prey per capita rate of change ($(1/N)(dN/dt)$). As usual, the optimal defense level, δ_{opt} is found by solving for δ in the equation $\partial w/\partial\delta = 0$, which in this case is rather straightforward:

$$\delta = ln\left(\frac{\theta qP}{1 + sN}\right)\frac{1}{\theta} \tag{3.22}$$

Not surprisingly, the optimal defense level is a function of both prey and predator density. We therefore note that if population densities are changing, δ_{opt} also changes accordingly. As predator density increases, so does the defense level. However, there is a negative effect of prey population on the optimal defense level, stemming from the negative density dependence implied by the decelerating functional response. The per capita risk of mortality from predation decreases with increasing prey density, due to a simple dilution effect, hence the benefit of further decreasing risk by investing in defense does not always outweigh costs.

The equilibrium density of the prey (N^*) and predator (P^*) populations is readily calculated by

$$N^* = \frac{d}{cqe^{-\theta\delta} - sd} \tag{3.23}$$

$$P^* = \left(\frac{1 + sN^*}{qe^{-\theta\delta}}\right)\left(r - \frac{rN^*}{K} - \delta\right) \tag{3.24}$$

Ives and Dobson (1987) have analyzed the stability conditions of this system. If $\delta^* = 0$ at equilibrium, then the system is stable if $s \leq 1/K$. If $\delta > 0$ at equilibrium, then the system is stable if $s \leq wr/K$.

Perhaps more interesting than stability properties per se is the indirect interaction between species of defended versus undefended prey. The first question is whether a system with undefended prey and predator can be invaded by a defended prey strategy; the second question is whether they can coexist. Suppose we have a system with an undefended prey species at equilibrium. A defended species can only invade the community if its fitness is positive when it is rare. If the system is at equilibrium and the intruding species is sufficiently rare, then dynamics are essentially governed by interactions between undefended prey and the predator. Since the intruding defended strategy is rare, its intrinsic per capita growth rate should be close to maximum, which we symbolize r_D. Consumption rates on both prey should be dependent on N_{ND}, the population density of the undefended prey species. When the other species are at equilibrium, the fitness of the rare defended species equals

$$w_D = r_D - \delta^* - \frac{qe^{-\theta\delta^*}P^*}{1 + sN^*_{ND}} \tag{3.25}$$

where $\delta^* = \ln(\theta q P^*/(1 + sN^*_{ND}))(1/\theta)$ and P^* and N^*_{ND} are given by equations 3.23 and 3.24. The invasibility criterion that equation 3.25 be greater than zero is not so easily analyzed. Figure 3.13 illustrates how the fitness of the defended species, when rare, varies with the defense efficiency parameter θ and the ratio between r_D and r_{ND}. If θ is large enough, then the defended species can invade even though $r_D < r_{ND}$. Hence, even though the defensive species pays some

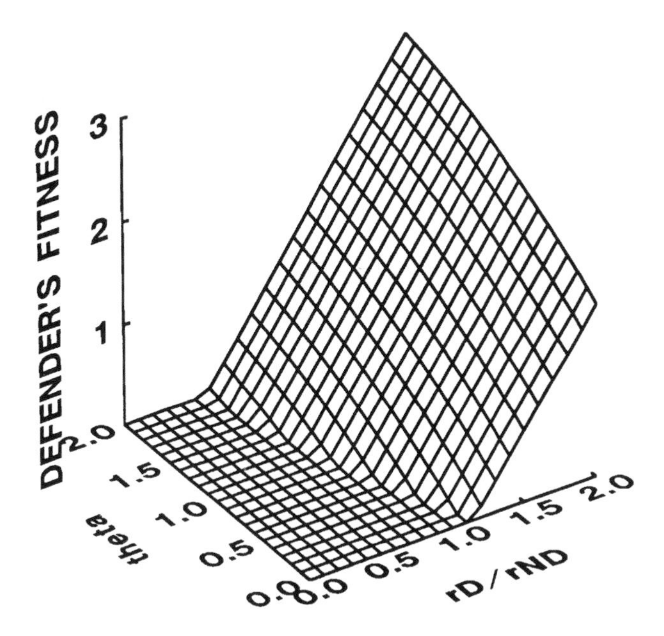

Figure 3.13. The fitness of the defended morph in relation to the defense parameter w and the ratio between the per capita rate of increase of the defended and the undefended morph in the population. The defended morph can increase when rare (positive fitness) if defense efficiency θ is large and if r of the defended morph is large in relation to the r for the undefended one.

physiological price for its defensive flexibility, it can still be an evolutionarily stable strategy (ESS). Only situations for which $\delta^* > 0$ and the equilibrium (N^*_{ND}, P^*) is locally stable are shown. Given that a defended species can invade, the next question is whether the two strategies can coexist. If we assume that the two prey species differ only in terms of defense investment (with the usual mortality price paid by the defender), we write the fully system

$$\frac{dN_D}{dt} = rN_D\left(1 - \frac{N_D}{K}\right) - \delta N_D - \frac{qe^{-\theta\delta}N_DP}{1 + s(N_d + N_{ND})} \tag{3.26}$$

$$\frac{dN_{ND}}{dt} = rN_{ND}\left(1 - \frac{N_{ND}}{K}\right) - \frac{qN_{ND}P}{1 + s(N_d + N_{ND})} \tag{3.27}$$

$$\frac{dP}{dt} = P\left(\frac{cq(N_De^{-\theta\delta} + N_{ND})}{1 + s(N_d + N_{ND})} - d\right) \tag{3.28}$$

Note that the two prey types do not compete for the same resources, they only interact through a common predator. The optimal defense level is calculated as previously, and we have $\delta^* = \ln((\theta qP)/(1 + sN_{ND} + sN_{ND}))(1/\theta)$. Handling this system analytically is rather cumbersome so we therefore simulate time dynamics over a range of parameter values. Figures 3.14–3.15 show comparisons of the two species separately, and when sharing the same predator, for a variety of parameter values of θ. The influence of defense on the time dynamics is clearly

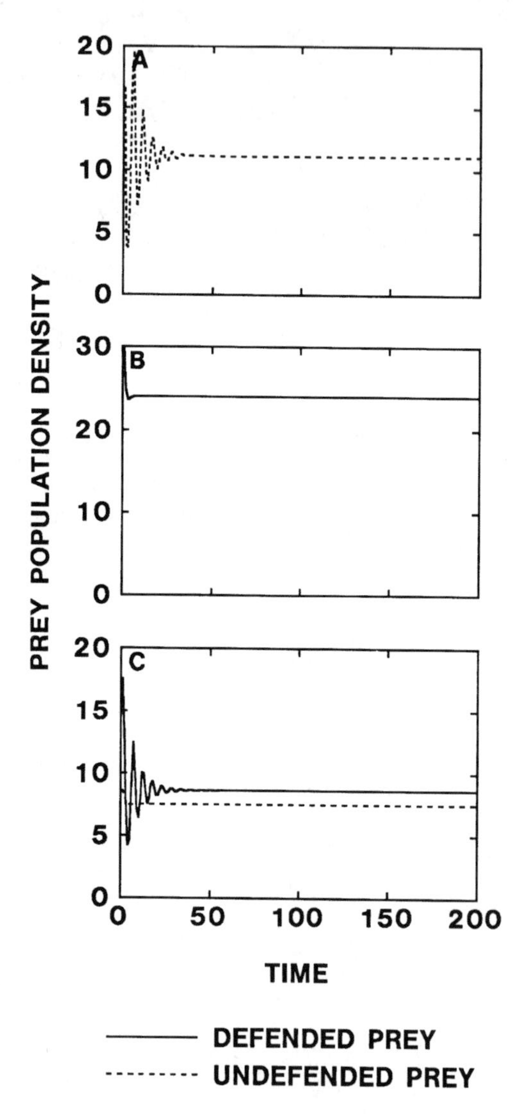

Figure 3.14. The time series of defended and undefended prey populations when the defense is very efficient ($\theta = 1.6$), according to equation 3.26–3.28. Parameter values, $s = 0.01$, $r = 2.0$, $c = 0.5$, $d = 1.0$, $q = 0.2$.

stabilizing when the defense is sufficiently efficient ($\theta = 1.6$, Fig. 3.14). At low carrying capacities and when defense is not very efficient ($\theta = 0.4$, Fig. 3.15), defense is not useful and the optimal level of defense is nil. If carrying capacity is high enough to push the system into the unstable region, then prey can turn defenses on and off at will, thereby reducing the predation rate when predator

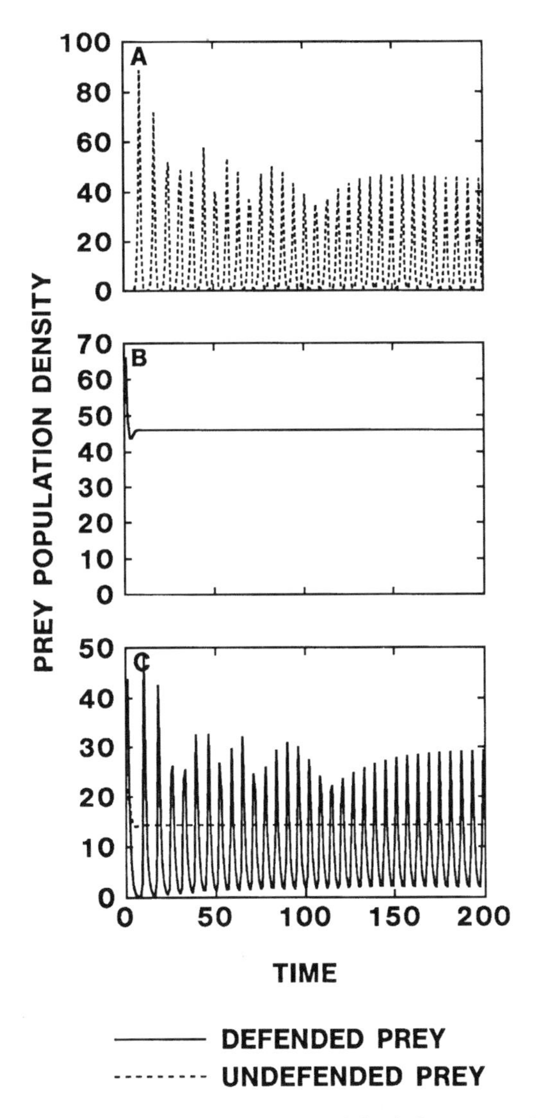

Figure 3.15. The time series of defended and undefended prey populations when the defense is not very efficient ($\theta = 0.4$) according to equations 3.26–3.28. Parameter values as in Fig. 3.14.

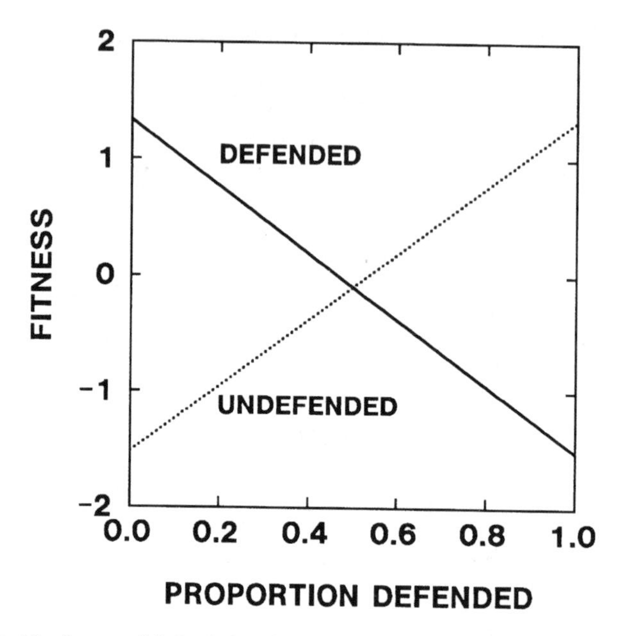

Figure 3.16. The fitness of defended and undefended morphs in relation to the proportion of defended individuals in the prey population. Fitness is negatively frequency dependent, potentially leading to stable coexistence between the two morphs.

densities are high, but growing fast when predator densities are low. Sharing a common predator gives the defended species an advantage and it can often vanquish the undefended prey species. The two prey types are consumed in proportion to their attack rates, so strictly speaking there is no diet choice by the predator, although diet composition certainly changes over time.

The relative success of the two prey species is illustrated in Figure 3.16. Here, the fitness of the two prey types is plotted against the proportion of the defended species in the total prey population. The figure shows that their respective fitnesses are negatively frequency dependent, a prerequisite for stable coexistence. The exact position of the intersection of the two fitness functions, if they indeed do intersect, depends on the optimal level of defense, which is determined in turn by the efficacy of defense and population densities at equilibrium (which are not independent). Whether the equilibrium proportion will be attained is also a matter of system stability. If the system parameters lead to instability, then mean population densities deviate from the equilibrium densities, often favoring competitive exclusion of one or the other genotypes, but not always.

3.5.2 Lagged Induced Defense

The foregoing treatment dealt with defense strategies that were instantaneously flexible. Foliage-feeding insects, for example, face such rapidly induced plant responses that time delays are immaterial (Nevonen and Haukioja 1991). This is not always the case. In particular, many plant defenses that are induced by grazing do not come into effect until long after the first attack has occurred (Karban and Myers 1989). It has been proposed that this may cause such delays in the response that it may help drive plant–herbivore oscillation (Haukioja and Hakala 1975; Karban and Myers 1989; Edelstein-Keshet and Rausher 1989; Lundberg et al. 1994).

We use the model given by equations 3.20 and 3.21 to illustrate this proposition. Instead of letting the optimal defense level be an immediate response to the actual population densities, we let prey adjust δ^* to the situation at time *t*, but the change in defense does not come into effect until time $t + 1$, implying a time lag in the onset of induced defense. It is not surprising that population variability increases if the optimal defense level is lagged one time step behind (Fig. 3.17).

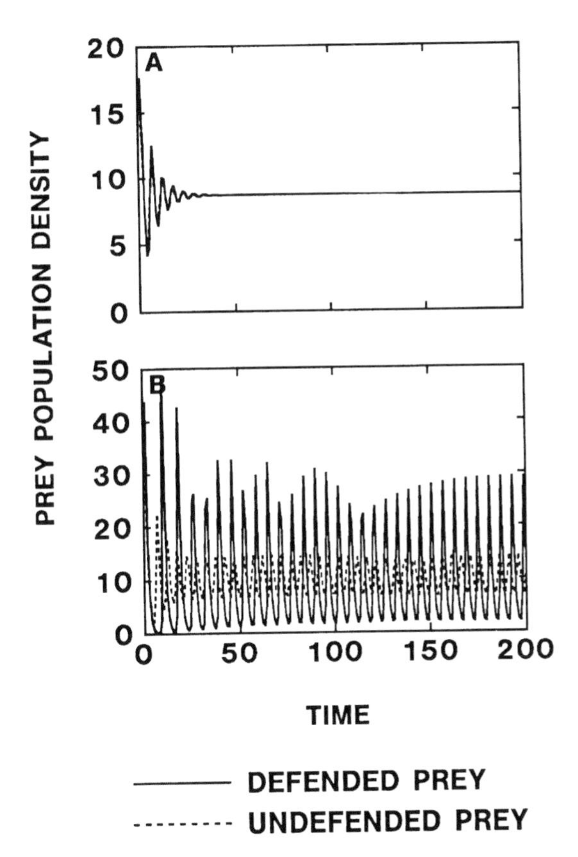

Figure 3.17. The time series of the prey population having perfectly timed (*A*) and time-delayed induced (*B*) defense. The induced defense leads to more variable (less stable) prey population dynamics. Parameter values as in Figure 3.14.

However, the optimal defense level will vary continually, trying to catch up with the changing population densities. In other models, e.g., Edelstein-Keshet and Rausher (1989), Adler and Karban (1994), and Lundberg et al. (1994), the defense does not have the foregoing lag structure. Instead, the induced defense has a predetermined decay rate after initiation. The dynamic effect is nevertheless similar: The lower the decay rate, the more delayed the population response.

3.6 SUMMARY

The risk of being eaten has led to a variety of evolutionary responses by prey, including avoidance of detection by predators and attack deterrence. In this chapter, we investigate the effects that behavioral and chemical defenses may have on predator–prey dynamics. Two major classes of defense traits are considered.

First, we let time allocation be a decision variable. We assume that prey individuals can either spend time foraging, positively influencing their rate of energy gain, or they can refrain from foraging, thereby avoiding predation but losing energy. A predator–prey system with prey individuals that optimally allocate their time between these two states is notably more stable than a system with constant prey activity. This stems from the fact that optimal time allocation creates a prey refuge that is particularly pronounced at high prey densities, causing positive density-dependent predation.

We extend this model to let the intermediate link in a 3-link food chain exhibit facultative time use. Here, the stabilizing effect is less pronounced, affecting carnivores and plants more than herbivores. In this system, herbivore feeding rates should be positively related to herbivore density. This ameliorative effect is due to the fact that as herbivore density increases, so does the carnivore threshold for a change in herbivore time allocation.

Second, we study a predator–prey system in which the prey population exhibits induced chemical defenses, affecting attack rates by predators. In addition to the stability conditions of this system, we also investigate the evolutionary stability of the chemical defense. We show that a defended species will be able to displace an undefended species under nontrivial conditions, but only if the defense is very efficient. We also show that the undefended and defended species can coexist when they share a common predator.

Most treatments of defenses do not account for the population dynamic aspects of the interaction between defended and undefended prey types (Augner 1995). That probably matters little if the systems in question are at dynamical equilibrium. However, once the predator–prey interaction is deterministically unstable (unstable equilibrium) or under significant stochastic influence and population densities vary significantly from the steady state, then dynamical considerations are inevitable. Only recently has the importance of stochastic predation risk been fully appreciated (Adler and Karban 1994; Åström and Lundberg 1994), still lacking the necessary combination of frequency and density dependence. Much progress has been made in understanding time allocation as a form of defense, both theoretically and empirically. This is less true of plant defenses or chemical or morphological defenses among animals. Far fewer attempts have been made to account for the frequency- and density-dependent effects and little theory is available on the dynamics of interactive populations. Finding good ways of testing the emerging theoretical frameworks should be high on the experimental agenda.

Finally, we discuss the dynamic effects of lagged defenses, i.e., defenses that only come into action after the prey has experienced predator attack. The time delay inherent in such defense mechanisms tends to destabilize trophic dynamics.

4 Habitat Use and Spatial Structure

Most ecological theory is founded on the dubious notion that environmental conditions are homogeneous. Although it is undeniable that environmental variability increases substantially with the spatial scale of observation (Bell et al. 1993), there is still ample environmental variation within $1 m^2$ plots (Beckett and Webster 1971; Tilman 1982; Lechowicz and Bell 1992). Hence, it is probably just as meaningful to think about spatial variation in topography, habitat, and resource distribution for a soil nematode as it is for a mountain gorilla.

Behavioral ecologists have long been interested in the effects of spatial heterogeneity in resources on patterns of consumer movement and patch use. In turn, population ecologists have devoted much recent theory to the dynamics of metapopulations in a fragmented landscape. Less attention has been devoted to a synthesis of these points of view. In this chapter, we consider the dynamical implications of fitness-maximizing movement strategies by both predators and their prey living in a heterogeneous landscape.

We consider two major aspects of environmental heterogeneity that bear on ecological dynamics: (1) ecological differences among habitats and (2) the spatial arrangement of patches of suitable habitat. To some degree, variation in both habitat quality and spatial structure characterize any real system. In order to disentangle the effects of habitat variability from the effects of spatial structure, however, we treat these issues separately.

In each case, we first develop a behavioral model that maximizes the rate of energy gain and apply it to a plant–herbivore system. We consider mechanistic competition among genotypes employing different strategies for habitat use. We then develop a slightly more complex behavioral model that optimizes the trade-off between energy gain and predation risk and apply this to a plant–herbivore–carnivore system. Carnivores and herbivores make optimal space use decisions in the 3-link food chains.

4.1 HABITAT VARIATION

Let us assume that the environment of some hypothetical organism consists of some areas of highly suitable habitat interleaved with areas of less suitable habitat. There are many possible behavioral responses to such environmental variability:

do nothing, emigrate periodically, or move whenever necessary to enhance fitness. If all else were equal, natural selection would always favor selection of the "best" habitat available. All else is seldom equal, however, in dynamic communities. Whether for reasons of intrinsic instability or external stochasticity, fitness is often changing over both space and time.

Population density often has a negative effect on individual fitness, either due to direct interference or resource depression. In a landmark paper, Fretwell and Lucas (1970) argued that the initial colonizers of a pristine environment ought to first concentrate in the best habitats. As population density within that habitat increases over time, intraspecific competition should inevitably lead to a localized decline in fitness. At some point, fitness in the best habitat would have eroded to the extent that individuals moving to intrinsically poorer, but as yet unpopulated, habitats would fare just as well as the original colonists in their crowded habitat. Provided that individuals are all of equal ability and knowledge, then natural selection would favor a spatial distribution such that all individuals, regardless of location, have identical fitness. This so-called "ideal-free distribution" favors individuals that move to alternate habitats when fitness gains within a given habitat fall below that in other habitats. Similar arguments can be applied to populations in which individuals differ in competitive ability or social status (Parker and Sutherland 1986; McNamara and Houston 1990; Korona 1990; Sutherland and Parker 1992). In this latter case, the distribution of individuals involves phenotypic-specific assessment of potential fitness, a game played against the rest of the population. We defer discussion of phenotype-specific strategies to Chapter 6, devoted to interference and territoriality.

An extensive body of theory has been shaped by this paradigm of "ideal" habitat selection (see Rosenzweig 1991 for an excellent introduction to this extensive and rapidly growing literature), with an impressive body of experimental tests of community structure and dynamics. Much of this work hinges on the simplifying assumption that population density is a useful index of fitness within habitats, which may not be valid for all communities. Most applications of density-dependent habitat theory have been in donor-controlled systems that ought to be well approximated by density-dependent models (Morris 1987, 1988, 1989; Abramsky et al. 1990, 1991; Brown et al. 1994). In interactive systems, however, fitness of consumers depends on both resource abundance at lower trophic levels and the risk of predation imposed by higher trophic levels, so fitness need not necessarily vary in straightforward fashion with consumer density. We need a more comprehensive set of models, based on ecological interactions across the food chain.

4.2 ENERGY-MAXIMIZING HABITAT USE

Assume that a predator feeds on only one species of prey that occurs in either of two habitats, but that the rate of effective search differs between habitats because of differences in vegetation cover or terrain. For example, one might be

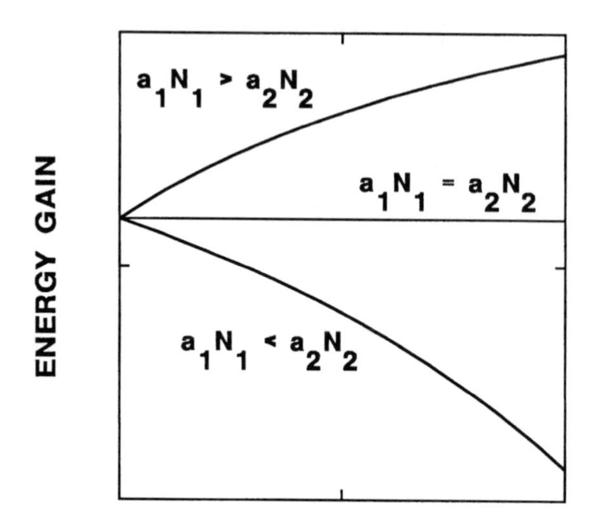

Figure 4.1. Rate of energy gain as a function of the proportion time spent in the better of 2 habitats, under three different ratios of prey abundance in the 2 habitats.

studying a predator whose efficiency of prey attack varies across vegetation communities, such as lions stalking antelope in areas of tall grass versus short grass (Van Orsdol 1984). If travel time between habitat patches is negligible, then the total time budget is composed of search time ($T_s = S_1 + S_2$ where $S_1 = \beta T_s$ and $S_2 = [1 - \beta]T_s$) and handling time in each habitat ($T_h = a_1\beta T_s hN_1 + a_2[1 - \beta]T_s hN_2$), where S_i = the time spent foraging in habitat *i*, β = the probability of foraging in habitat 1 and $1 - \beta$ = the probability of foraging in habitat 2 (Lawlor and Maynard Smith 1976; Murdoch 1977; Rosenzweig 1981; Hubbard et al. 1982). According to this scenario, time spent searching for prey in habitat 1 must be inversely related to the time devoted to feeding in habitat 2. The rate of energy gain can be calculated from the following equation:

$$\phi = \frac{\beta e a_1 N_1 + (1 - \beta) e a_2 N_2}{1 + \beta a_1 h N_1 + (1 - \beta) a_2 h N_2} \tag{4.1}$$

This function is either monotonically increasing or decreasing with respect to β (and therefore also $1 - \beta$), but is never maximized at intermediate values (Fig. 4.1). Hence, the optimal strategy is to exclusively use habitat 1 or else use habitat 2, but never use a mixture of both habitats. The optimal choice of habitat depends on the densities of prey in each habitat and the rates of search in each habitat. Energy gain is equal in each habitat when

$$\frac{ea_1N_1}{1 + a_1hN_1} = \frac{ea_2N_2}{1 + a_2hN_2} \tag{4.2}$$

which reduces algebraically to the following equation

$$a_1N_1 = a_2N_2 \tag{4.3}$$

The behavioral implications imposed by spatial segregation of prey are (1) that an adaptive predator should search exclusively in one habitat or the other at any given time, but never both and (2) that the optimal choice of habitat depends on the densities of prey in both.

It is unrealistic to suppose that habitat selection would ever be perfect. If we assume that the probability of a "mistake" is small when $a_1N_1 >> a_2N_2$ or $a_1N_1 << a_2N_2$, but that mistakes are more likely when $a_1N_1 \approx a_2N_2$, then the probability of habitat use can be modeled according to a sigmoid function (Nisbet et al. 1993; Fryxell 1997):

$$\beta = \frac{e^{zN_1}}{e^{zN_1} + e^{za_2N_2/a_1}} \tag{4.4}$$

As in our previous sigmoid behavioral switches, this function ranges in value between 0 and 1, with an inflection at the ratio of prey densities in the 2 habitats at which $a_1N_1 = a_2N_2$. The parameter z controls the rate at which the curve saturates on either side of this threshold.

For an indiscriminate forager, the proportion of attacks occurring in habitat i plotted against the proportion of prey living in habitat i should fall on a straight line of slope = 1, provided that the forager perceives and attacks each prey with equal probability. On the other hand, if predators choose facultatively between habitats in order to maximize energy gain (Murdoch 1977; Hubbard et al. 1982), then the best solution is for the forager to preferentially attack prey in habitat 1 when $N_1/N_2 >> a_2/a_1$, but preferentially attack prey in habitat 2 when $N_1/N_2 << a_2/a_1$. Hence, a scatterplot of habitat electivity versus relative availability in the two habitats should have a sigmoid shape, with values of 1 above the critical threshold and values of 0 below the threshold (Fig. 4.2; Murdoch 1977; Hubbard et al. 1982).

As a result of this sigmoid probability of habitat use, the functional response by a predator exercising facultative habitat selection should depend on densities of prey in both habitats and there should be an inflection near the point of habitat switching (Fig. 4.3; Murdoch 1977; Hubbard et al. 1982; Holt 1983). Incidentally, this circumstance is one of the few in which classic prey switching proves to be advantageous in an evolutionary sense (Holt 1983), although there are certainly other adaptive causes of sigmoid functional responses (Abrams 1982, 1987*a;* Sih 1984; Mitchell and Brown 1990; Fryxell and Lundberg 1994). The rate of

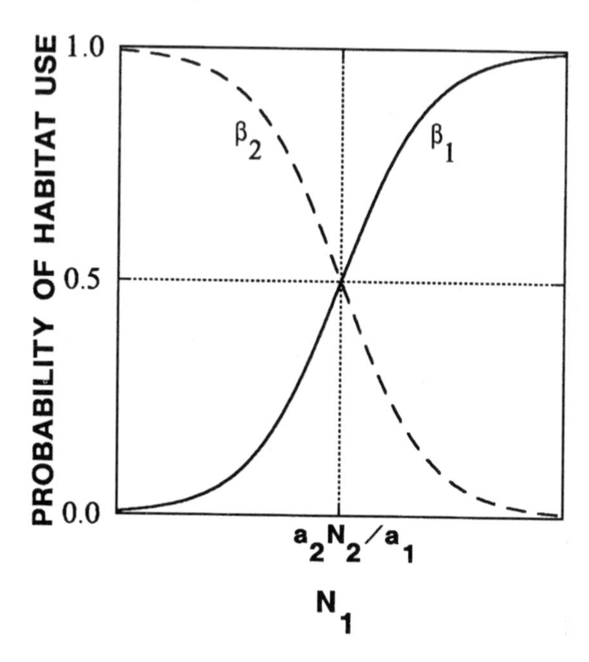

Figure 4.2. Probability of use of 2 potential habitats (β_1 and β_2) by a forager using facultative habitat choice to increase the rate of energy gain.

predation is accelerating for all prey densities below the critical threshold (i.e., when $0 < N_1 < a_2N_2/a_1$). Hence, there is a broad range of prey densities over which the per capita risk of mortality is density dependent, and therefore potentially stabilizing (Murdoch and Oaten 1975; Murdoch 1977; Sih 1984; Fryxell and Lundberg 1993).

4.2.1 Experimental Evidence for Optimal Habitat Choice

Perhaps the most straightforward experiment to test the basic mechanism of optimal habitat selection is simply to manipulate the proportions of prey in different habitats. One of the earliest and best examples of such a prey selection experiment was conducted by Murdoch et al. (1975). Initial lab trials with naive guppies showed that individual fish tended to have strong preferences (approaching 100%) for either tubificid worms at the bottom of an aquarium or fruitflies floating on the water surface. Murdoch and coworkers then altered the ratio of tubeworms versus fruitflies to test whether the fish accordingly changed their feeding preferences. The data clearly demonstrated that fish reversed their habitat preferences as prey density in the alternate habitat was augmented (Fig. 4.4), just as predicted by the model of facultative habitat selection. This case is particularly compelling because the spatial segregation of prey is without question.

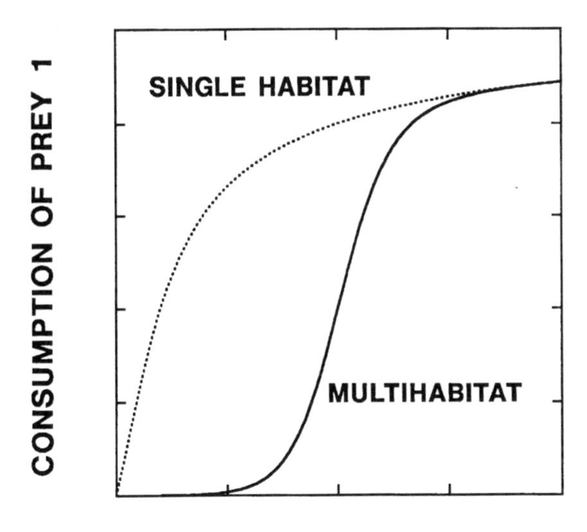

Figure 4.3. Functional responses of a forager specializing on prey in the open habitat compared to a forager facultatively choosing between prey in 2 habitats to increase energy gain. The rate of consumption of prey 1 in the open habitat in each case is shown. Note the inflection in the functional response surface at the point of switching from the 1 habitat to the other.

Perhaps more importantly for our purposes, Murdoch et al. (1975) also observed accelerating functional responses to changes in prey density (Fig. 4.5), as predicted by the optimal habitat selection model (Fig. 4.3).

Murdoch et al.'s (1975) experiment involved two species of prey in two separate habitats. There is similarly strong evidence for facultative habitat selection by predators presented with a single species of prey in two habitats. For example, Milinski (1979) provided sticklebacks at either end of an aquarium with food delivered at constant rates. The experiment consisted of altering the rates of food delivery, i.e., altering habitat quality, and measuring the pattern of spatial distribution relative to that expected according to ideal free principles. His results clearly demonstrated that the fish shifted their distribution in response to changes in food delivery. For example, a food delivery ratio of 2:1 was matched by a 2:1 ratio in fish densities at the two ends of the tank. When the ratio was shifted or even reversed, the fish tracked changes in resource distribution. It is interesting to note, however, that the fish were not equal in fitness: Invariably some fish obtained higher reward rates than others, presumably due to their dominance status. Nonetheless, the fitness-maximizing spatial distribution was upheld.

A less direct way to test the ideal free model is simply to record changes in

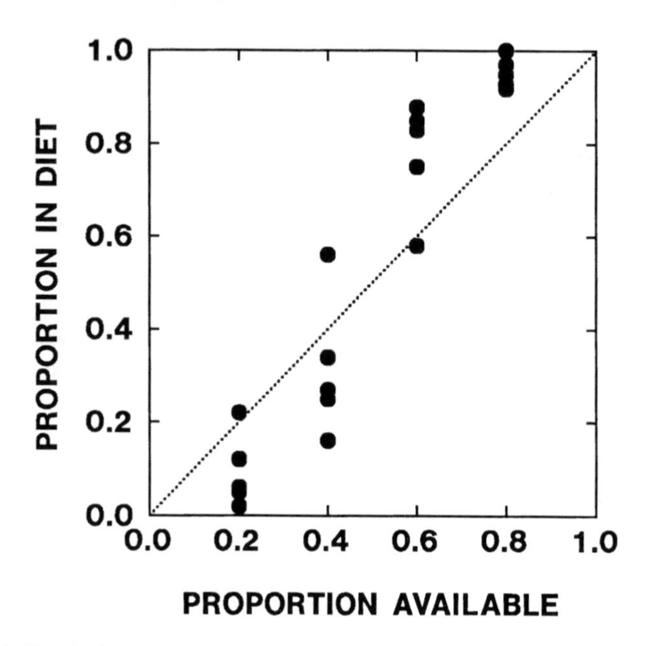

Figure 4.4. Facultative microhabitat switching by guppies in the laboratory (redrawn from Murdoch et al. 1975), as demonstrated by the proportion of tubificid worms in the diet (p_1) versus the proportion in the environment (F_1). The dotted line indicates the expectation if guppies used both habitats (and therefore both prey) with equal probability.

the proportion of consumers in each habitat as consumer density changes over time. If individuals assess within-habitat fitness in relation to their expectation from other habitats, then one might predict that habitat preferences should be most extreme when consumers are rare and most uniform when consumer density is high (and resource density in the preferred habitat is reduced). Such evidence for habitat preferences is clearly demonstrated by Morista's (1952) classic study of habitat preferences of antlions or Rosenzweig and his colleagues' (Pimm et al. 1985; Rosenzweig 1986) studies of feeder use by hummingbirds. In each case, as consumer density increased, the relative use of preferred habitats declined whereas that of poorer habitats increased.

4.2.2 Habitat Choice and Predator–Prey Dynamics

The rate of change of prey over time in habitat *i* depends on the rate of consumption multiplied by predator density and the probability of search in habitat *i* (Nisbet et al. 1993; Fryxell and Lundberg 1993):

$$\frac{dN_1}{dt} = rN_1\left(1 - \frac{N_1}{K_1}\right) - \frac{\beta a_1 N_1 P}{1 + \beta a_1 h N_1 + (1 - \beta) a_2 h N_2} \tag{4.5}$$

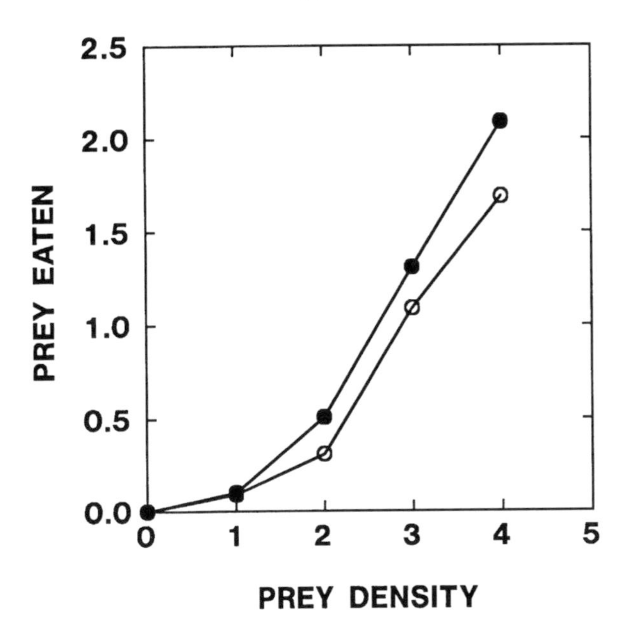

Figure 4.5. The functional response of Murdoch et al.'s (1975) guppies in relation to prey density. The inflection in the functional response for each prey type was apparently due to microhabitat switching, much like the predictions of the facultative habitat use model.

$$\frac{dN_2}{dt} = rN_2\left(1 - \frac{N_2}{K_2}\right) - \frac{(1 - \beta)a_2N_2P}{1 + \beta a_1hN_1 + (1 - \beta)a_2hN_2} \tag{4.6}$$

$$\frac{dP}{dt} = P\left(\frac{ce[\beta a_1N_1 + (1 - \beta)a_2N_2]}{1 + \beta a_1hN_1 + (1 - \beta)a_2hN_2} - d\right) \tag{4.7}$$

In order to obtain a perspective on the effects of energy-maximizing habitat use, we compare our results to alternate models in which β is constant, implying that at any given moment a constant fraction of foragers is switching habitats. In these simulations we assumed a single sedentary resource of profitability $e/h = 2$, found in two habitats differing in cover such that the rate of effective search in habitat 1 exceeds that in habitat 2 ($a_1 > a_2$), but otherwise similar in all respects (K, r, and d identical in each habitat).

The range of stable parameter combinations for a specialist forager in habitat 1 is smaller than the range of stable combinations for a specialist forager in habitat 2 (Fig. 4.6*A,B*). This is not surprising, given that $a_1 > a_2$, implying that foragers are more efficient in acquiring resources in habitat 1. A coupled system with constant probability of emigration has an identical range of stable parameter combinations as habitat 1 in isolation (Fig. 4.6*C*). A coupled system with faculta-

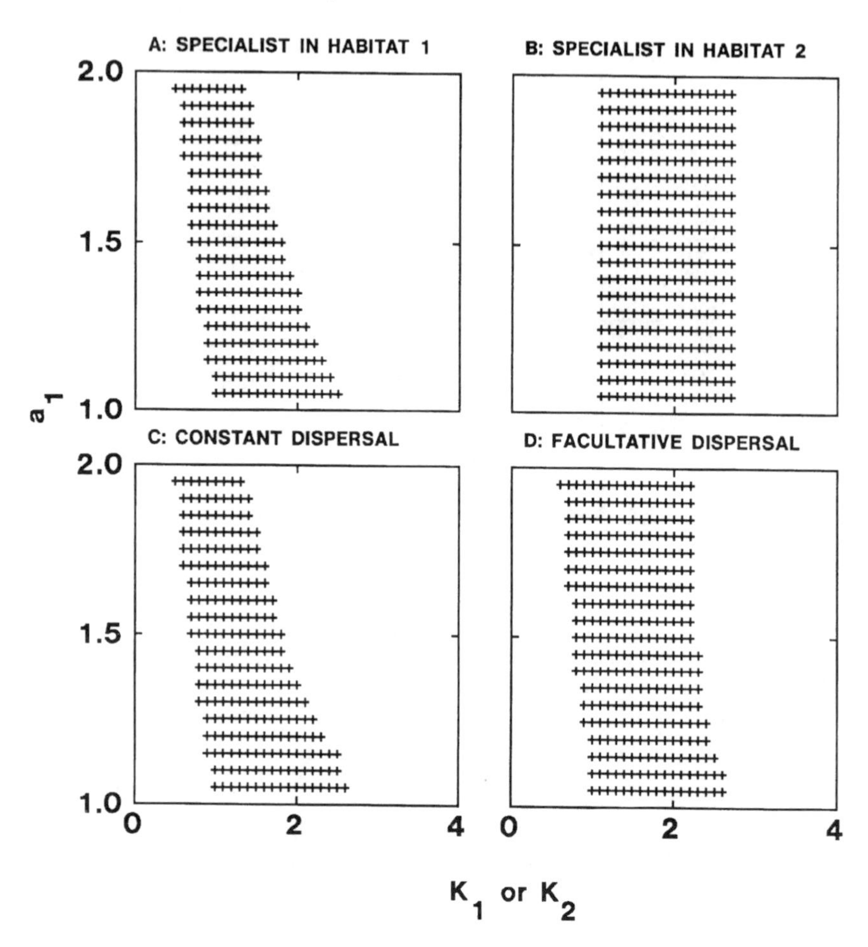

Figure 4.6. Parameter combinations (carrying capacity *K* versus search rate a_1 in the open habitat) yielding stable interactions between stationary prey in either of 2 habitats and mobile predators. In *A* and *B*, predators never disperse from their habitat. Predators have a constant probability of dispersal (5%) in *C* and distribute themselves in an ideal free distribution in *D*, following an energy-maximizing decision rule. Note that the stability range for the facultative predator is similar to the combined ranges of the stationary predators in either habitat. Stability in the constant dispersal system is identical to that of the least stable habitat. The following parameter values were used: $a_1 = e = 2$, $a_2 = h = c = d = r = 1$, $z = 4$.

tive dispersal, on the other hand, has a wider range of stable parameter combinations, merging the stable ranges of both habitat 1 and 2 (Fig. 4.6*D*).

To clarify these conclusions, we illustrate a typical case in which $a_1 = 2$ (an open habitat) and $a_2 = 1$ (a dense habitat). The increased efficiency of search by specialist foragers in habitat 1 leads to oscillations (Fig. 4.7*A*), unlike the stable conditions experienced by specialist foragers in habitat 2 (Fig. 4.7*B*), for the

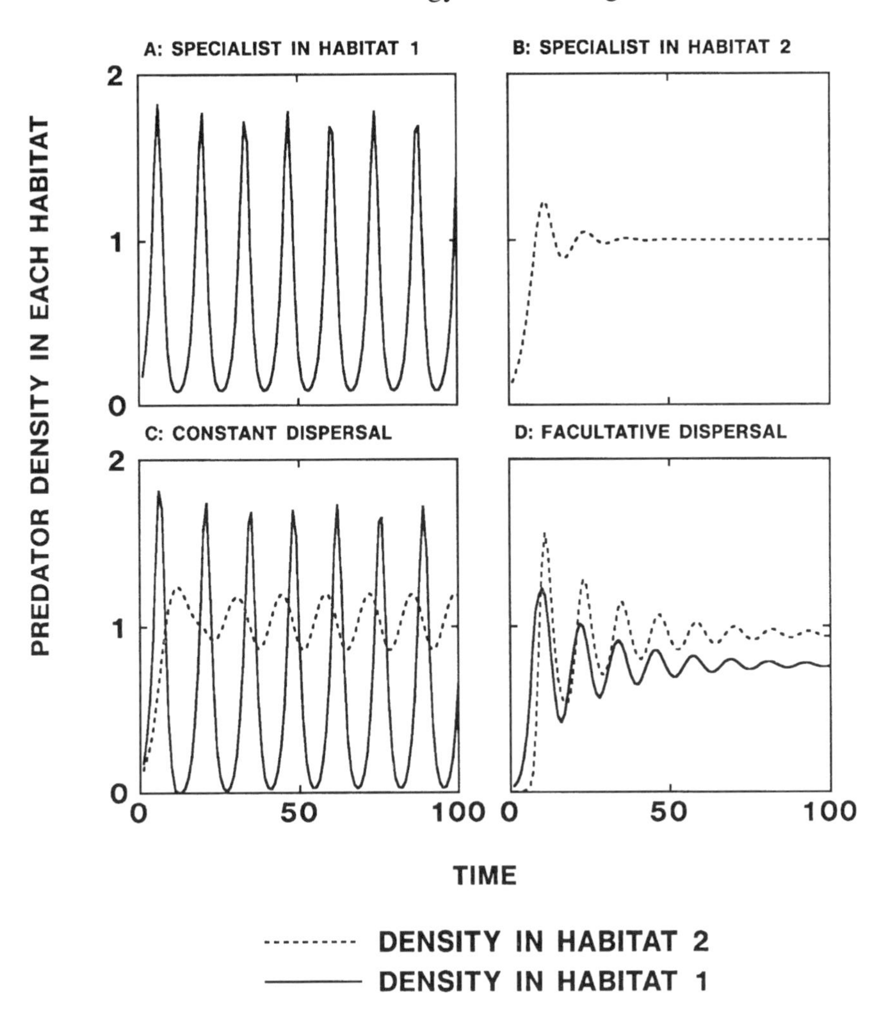

Figure 4.7. Predator population dynamics over time according to three behavioral scenarios (*A* and *B*, no dispersal; *C*, constant dispersal probability of 5%; and *D*, facultative dispersal according to an energy-maximizing decision rule. The following (unstable) parameter values were used: $a_1 = e_1 = e_2 = 2$, $a_2 = h_1 = h_2 = c = d = r = 1$, $z = 4$, $K_1 = K_2 = 2$.

reasons discussed in Chapter 1. Forager movements synchronize dynamics among habitats (Fig. 4.7*C,D*), often leading to stability of the system with a facultative forager (Fig. 4.7*D*).

Facultative habitat use confers a stabilizing influence because the probability of resource use within a given habitat sharply drops as resource availability within that habitat declines relative to alternative habitats. In the vicinity of the point of habitat shift, the per capita risk of prey mortality changes from being a negative, destabilizing function of resource density to a positive, stabilizing function of prey density. When one habitat is intrinsically stable, but the other is intrinsically unstable, the habitat switching point is near the equilibrium point for predators, at which reproduction (the numerical response) just balances mortality. Once again, pronounced variability around the habitat switching point increases stability: Behavioral variability among individual foragers increases community stability. Krivan (1997) found a similar stabilizing effect in pure Lotka-Volterra systems with optimal habitat choice by consumers.

The density of facultative foragers is often higher in the dense habitat than the open habitat (Fig. 4.7*D*), which seems counterintuitive. The explanation stems from the stabilizing effect of switching. Per capita mortality is positively density dependent when prey density in habitat 1 is slightly less than the switch point, that is N_1 is slightly less than a_2N_2/a_1 (Fig. 4.4). As a result, the proportion of foragers emigrating from the open habitat exceeds that emigrating from the dense habitat, leading to an imbalance in predator densities in the two habitats. This has troublesome implications. Community ecologists often use population density as an index of habitat suitability or quality, yet our nonlinear models would suggest that the converse might sometimes be more appropriate.

We have only considered the case of a single prey species distributed among two different habitats. With slight modification, our model would be just as relevant to a forager faced with different prey species that are spatially segregated. For example, migratory wildebeest and zebra commonly join together during their seasonal movements to and from the Serengeti Plains in Africa. Transects across the vast migratory herds reveals a well-defined pattern of localized spatial segregation between wildebeest and zebra (Sinclair 1985). Hence, at the perceptual scale appropriate to a lion, there might well be dynamically relevant patterns of spatial segregation occurring among alternate prey species.

4.2.3 Habitat Choice and Competitive Dynamics

Alternative patterns of habitat use by potential competitors can lead to a variety of evolutionarily stable community structures (Lawlor and Maynard-Smith 1976; Rosenzweig 1981; Matsuda and Namba 1989; Brown 1990; Brown and Vincent 1992), the precise makeup of which depends on the scale of habitat heterogeneity, costliness of alternate patterns of habitat use, and the rate of gene flow between habitats (Brown 1986, 1990, 1995; Brown and Pavlovic 1992). A common outcome of these models is that community richness of competitors is limited by the number of different resource "types" available, regardless of whether those "types" refer to different species or a single species of resource distributed across different habitats (Matsuda and Namba 1989; Brown 1990; Brown and Pavlovic 1992; Brown and Vincent 1992). Hence, a single habitat or resource tends to select for one consumer strategy, two prey types beget two consumer strategies, etc. Most of these models are highly stable, so they may have limited applicability to unstable communities, nor can they tell us much a priori about the effect of mechanistic competition on community dynamics.

We systematically examined constraints on competitor community structure through simulation of four consumer strategies: pure opportunistic foraging (equal probability of foraging in either of 2 habitats at any point in time), facultative foraging (probability of foraging in a given habitat dictated by the energy maximizing strategy embodied in eqn. 4.2), a specialist foraging in an open habitat 1, or a specialist foraging in a dense habitat 2. Habitat quality was indexed by the area searched per unit search time (a), reflecting habitat-specific variation in cover or topography (Brown 1988). We initiated each simulation at the carrying capacity of each resource and low, equal densities of each of 4 forager morphs. We then simulated temporal dynamics according to the following equations specifying changes in the 4 foragers using resources in 2 habitats:

$$\frac{dN_1}{dt} = rN_1\left(1 - \frac{N_1}{K_1}\right) - \sum_{i=1}^{4} \frac{\beta_i a_1 N_1 P_i}{1 + \beta_i a_1 h N_1 + (1 - \beta_i) a_2 h N_2} \tag{4.8}$$

$$\frac{dN_2}{dt} = rN_2\left(1 - \frac{N_2}{K_2}\right) - \sum_{i=1}^{4} \frac{(1 - \beta_i) a_2 N_2 P_i}{1 + \beta_i a_1 h N_1 + (1 - \beta_i) a_2 h N_2} \tag{4.9}$$

$$\frac{dP_i}{dt} = P_i\left(\frac{ce[\beta_i a_1 N_1 + (1 - \beta_i) a_2 N_2]}{1 + \beta_i a_1 h N_1 + (1 - \beta_i) a_2 h N_2} - d\right) \quad for\ i = 1 - 4 \tag{4.10}$$

Our results suggest that an evolutionarily persistent community can be composed of 1, 2, 3, or even all 4 strategies, with the outcome dependent on ecological parameters of the system (Fig. 4.8). As in other consumer-resource models, stability in the habitat-structured models is inversely related to resource carrying capacity. Interestingly, there are a wide variety of parameter combinations (asterisks in Fig. 4.8) in which all competitors can survive. In our discrete habitat system, there are only two resource types, yet the system can readily support 3 or even 4 competitors.

It is well established that dynamical instability or analogous forms of environmental stochasticity can prevent competitive exclusion in simple systems with 1 prey species and 2 predators (Koch 1974; Levins 1979; Armstrong and McGehee 1980). Recent theoretical work of a similar nature by Wilson and Yoshimura suggests that environmental stochasticity can prevent competitive exclusion, particularly when one of the competitors employs facultative habitat selection (Wilson and Yoshimura 1994). Instability per se is not essential for a rich competitor community in our system, however, because the richest communities are locally stable (Fig. 4.9).

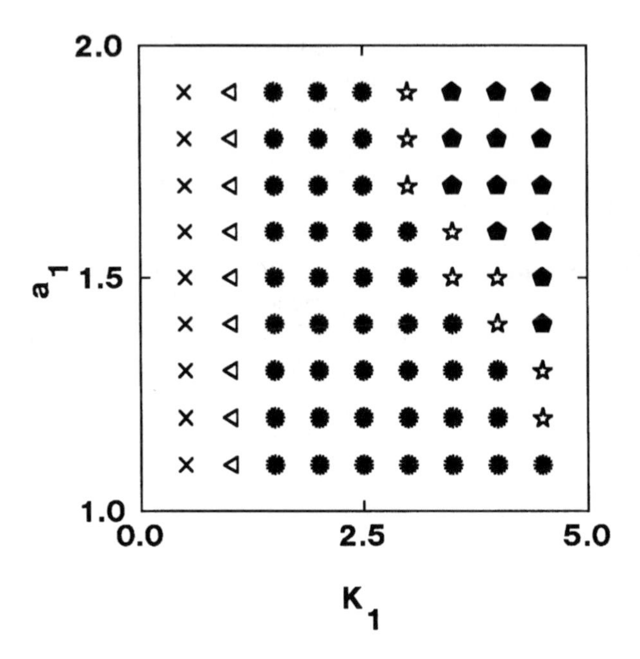

Figure 4.8. Evolutionarily persistent ensembles of competitors using different foraging strategies in relation to environmental parameters (carrying capacity K versus search rate a_1 in the open habitat). The following parameter values were used: $e = 2$, $a_2 = h = c = d = r = 1$, $z = 4$, $K_2 = K_1$. Symbols correspond to the following ensembles: crosses, no strategy can persist; triangles, the specialist in the open habitat 1 persists; asterisk, opportunist, facultative, and specialists in both habitats persist; stars, opportunist, facultative, and the specialist in habitat 2 persist; pentagons, facultative and the specialist in habitat 2 persist.

The perpetuation of several competitors in the same system requires that average fitness of each competitor must be equal over time. For example, sometimes the specialist has the highest fitness, but at other times it has the lowest fitness. Other strategies vary more moderately, but always in symmetrical fashion: Periods of high relative fitness are always counterbalanced by periods of poor fitness relative to other competitors. As long as the system favors such counterbalancing effects, diversity is enhanced.

One can readily evaluate the combinations of N_1 and N_2 at which each competitor would have a rate of growth equal to 0 (the zero growth isoclines first discussed in Chapter 1). Zero growth isoclines for each of the forager strategies have a common point of intersection in the N_1-N_2 phase-plane (Fig. 4.10). As long as that equilibrium is stable, each of the strategies have equal fitness, so all of them can coexist. Hence, competitive coexistence in our habitat-structured models is due to the facultative predator stabilizing the community, with an equilibrium at which the fitness landscape of all competitors is flat.

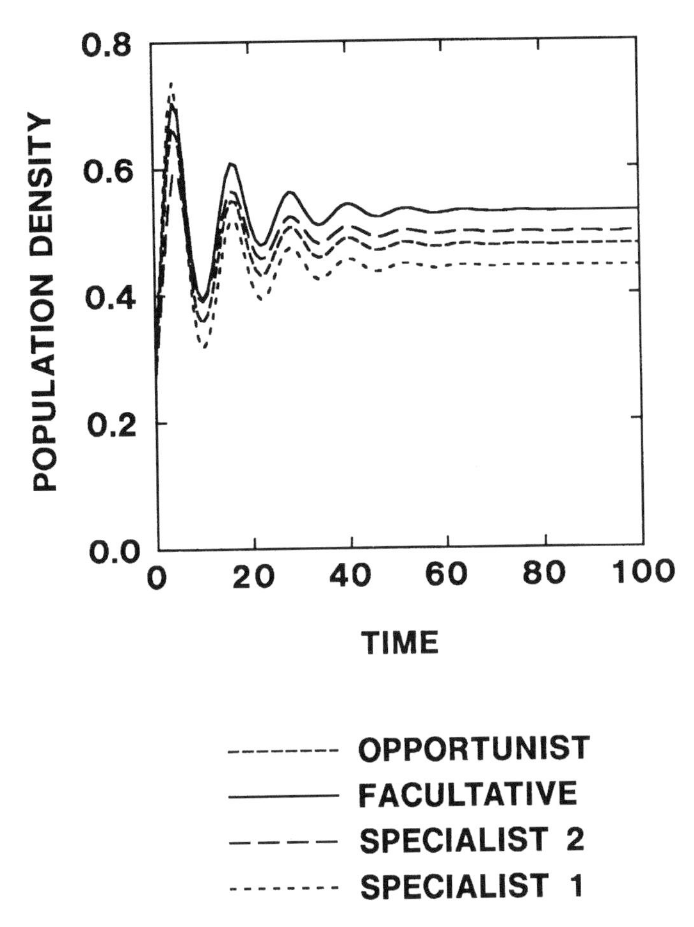

Figure 4.9. Population dynamics over time for a coexisting ensemble of competing consumers. The following parameter values were used in the simulation: $K_1 = K_2 = 2$, $a_1 = 1.3$, $e = 2$, $a_2 = h = c = d = r = 1$, $z = 4$.

4.3 EVASION OF PREDATORS BY PREY

It would also benefit mobile prey to take advantage of asymmetries in habitat structure. If all else were equal, one would expect prey individuals simply to choose habitats with the lowest risk of predation. However, it is inevitable that concentrated use of safe habitats would reduce access to resources or increase the risk of mortality from disease or intraspecific aggression. Hence, there may be ecologically relevant trade-offs between costs and benefits in each habitat potentially available (Sih 1984).

Numerous models have been proposed for predicting evolutionarily stable strategies for predator avoidance (e.g., Sih 1984; Mangel and Clark 1986; Gilliam

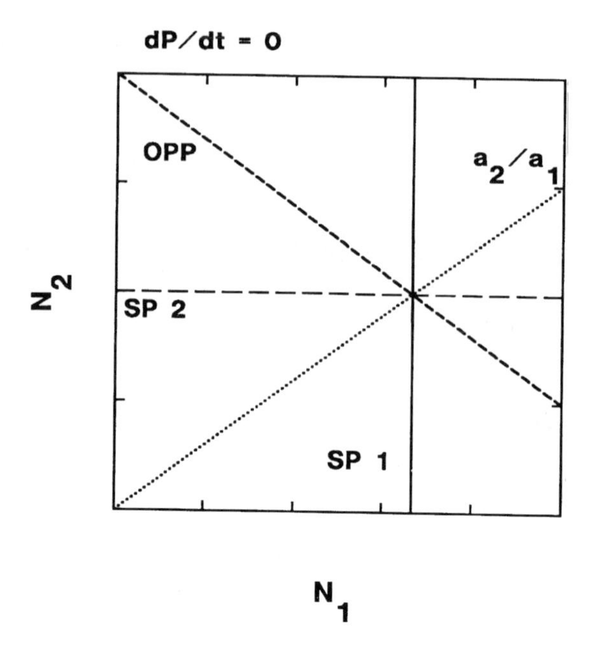

Figure 4.10. Zero growth isoclines for opportunists, habitat 1 specialists, and habitat 2 specialists, as well as the switching point (a_2/a_1) for a facultative predator. Note that all strategies have equal fitness at equilibrium.

and Fraser 1987; Brown 1992; Abrams and Matsuda 1993; Houston et al. 1993). Most arrive at similar predictions—not surprising given the similarities in structure. If reproductive gains and predation risk are positively associated, then the optimal strategy will involve partitioning habitat use such that the marginal loss of fitness due to predation equals the marginal fitness gain due to improved resource access.

For convenience, let us focus first on a herbivore, in the middle of a 3-species food chain. We will later add a carnivore that also chooses among habitats in order to maximize fitness. The first component of herbivore fitness is the per capita rate of recruitment of offspring, which equals the herbivore functional response multiplied by conversion efficiency (c) and energy content (e):

$$\phi_i = \frac{cea_iR_i}{1 + a_ihR_i} \tag{4.11}$$

where a_i = the area searched per unit time by a herbivore in habitat i, h = the handling time of each plant harvested by the herbivore, and R_i is the density of plants in habitat i. The fitness cost to a herbivore in habitat i is composed of the carnivore functional response divided by herbivore density plus the density-independent rate of herbivore mortality (d):

$$\theta_i = \frac{A_i P}{1 + A_i H N_i} + d \tag{4.12}$$

where A_i = the area searched per unit time by a carnivore in habitat i, H = the handling time per herbivore ingested, and N_i = herbivore density in habitat i.

One can now estimate fitness by subtracting the per capita risk of mortality from the per capita rate of recruitment due to foraging ($w_i = \phi_i - \theta_i$). It can be readily verified that herbivore fitness in habitat i is negatively related to carnivore density in habitat i ($\partial w_i / \partial P_i < 0$), but positively related to both plant density ($\partial w_i / \partial R_i > 0$) and the density of other herbivores ($\partial w_i / \partial N_i > 0$). In the latter case, fitness is enhanced by other herbivores via simple dilution of predation risk, due to the fact that the carnivore functional response is monotonically decelerating. Both $\partial w_i / \partial R_i$ and $\partial w_i / \partial N_i$ are asymptotic nonlinear functions, because they stem from functional responses of similar shape.

If one habitat yields a higher rate of resource use and is also safer from predators, then there is no behavioral trade-off of dynamical importance: Foragers should simply concentrate in the better habitat until resources are greatly depleted. It's more interesting to consider the situation in which a herbivore must balance the benefits of feeding in an environment yielding high rates of foraging (a large) but costing in turn increased risk of predation by carnivores (A large). This scenario is particularly likely when rates of search for both herbivores and carnivores are influenced by vegetation cover or physical relief: Habitat-specific effects are likely to apply similarly to different links in the food chain.

When plant densities are high, predators are common, or other herbivores are rare, the optimal habitat distribution for herbivores should tend to favor safe habitats. Conversely, when plant densities are low, herbivore densities are high, or predator densities are low, then the potential benefits of improved recruitment obtained by foraging in the better habitat might well supersede the increased risk of mortality. The model predicts that herbivore preference for safer habitats should decline with increasing herbivore or plant density or increase with increasing carnivore density.

One simple way to represent facultative choice between habitats is to model the probability of habitat use in relation to relative fitness gains. We once again assume that some variation is to be expected around the optimal step function:

$$\beta = \frac{e^{zw_1}}{e^{zw_1} + e^{zw_2}} \tag{4.13}$$

According to this formulation, the herbivore would tend to use habitat 1 exclusively when $w_1 >> w_2$, switch when $w_1 = w_2$, and exclusively use habitat 2 when $w_2 >> w_1$.

One can use the same logic applied earlier (eqns. 4.1–4.4) to predict carnivore fitness as a balance between per capita reproduction due to energy intake and per capita mortality:

$$W_i = \frac{CEA_iN_i}{1 + A_iHN_i} - D \tag{4.14}$$

where C = the conversion efficiency of carnivores, E = the energy content of each herbivore ingested by a carnivore, and D = the per capita mortality rate of carnivores. Once again we can use a sigmoid function to predict the probability of use of each habitat:

$$\gamma = \frac{e^{zW_1}}{e^{zW_1} + e^{zW_2}} \tag{4.15}$$

where W_i = predator fitness in habitat i and γ_i = the probability that carnivores remain in habitat i. As shown earlier, this model predicts that carnivores would shift from one habitat to the other when energetic rate of gain in the first habitat exceeds that in the second.

4.3.1 Experimental Evidence of Predator Evasion

An extensive body of evidence corresponds to the predicted risk-sensitive pattern of habitat use by prey species (see review by Lima and Dill 1990). Perhaps one of the earliest and strongest demonstrations that habitat use strategies reflect a delicate balance between the need to eat and avoid being eaten comes from Sih's (1980) laboratory experiments with the predatory aquatic insect *Notonecta*. In this particular case, late instar larvae cannibalize small instar larvae. Sih created artificial habitats with high densities of both cannibals and fruitflies or low densities of fruitflies and cannibals. The observed pattern of habitat use by different *Notonecta* instars reflected the trade-off between energy gain and predation risk. Large larvae, which are impervious to cannibalism tended to concentrate in the resource-rich habitat whereas small instars tended to select the safe, but resource-poor habitat. The degree of selectivity was directly associated with net fitness.

An even more precise demonstration of the balancing of energetic gains and predation risk comes from Nonacs and Dill's (1990) work with social ant colonies (*Lassius pallitaris*). Through an elegant series of titration experiments, Nonacs and Dill manipulated the ratio of reward rates to foraging ants in safe versus risky laboratory habitats. As predicted, use of the risky habitat increased with the relative rate of energetic reward. Under all experimental conditions, however, the ants always preferred the safe habitat to the risky one, which is predicted by essentially all marginal fitness balancing models (Gilliam and Fraser 1987; Brown 1992; Houston et al. 1993).

Such laboratory experiments offer elegant and tightly controlled experimental designs. It is less obvious whether habitat selection can be as readily predicted in the messy real world, where habitat quality is intrinsically more variable and difficult to measure. Perhaps the best demonstration that relatively simple models of habitat assessment can be usefully applied under realistic circumstances is the long-term study by Werner and coworkers on habitat use by bluegill sunfish (*Lepomis macrochirus*) in experimental ponds and small lakes in Michigan. They found that there were strong seasonal changes in the rates of energy gain obtainable within habitat categories (open water, vegetation, and sediments) and that the rankings of these habitats changed throughout the year. In the absence of predators, sunfish shifted habitats in accordance with the seasonal changes in rates of energy gain (Werner et al. 1983a). In the presence of a predator, vulnerable younger sunfish fed in energetically suboptimal habitats, whereas older invulnerable sunfish continued to choose the habitat offering the highest rate of energy gain (Werner et al. 1983b). Field observations in a number of lakes verified that ontogenetic habitat shifts were associated with changes in the relative importance of predation risk versus energy gain on fitness: Sunfish fry are at little risk of predation and forage in the open, juvenile fish are at risk and consequently forage in the littoral zone, whereas invulnerable older fish return to the open water (Werner and Hall 1988).

Equally convincing are a series of simple, but ingenious, field experiments conducted on granivores by J. S. Brown and coworkers. The protocol in each case involves using trays with a premeasured quantity of seeds mixed into sand as controlled surrogates for surrounding habitats. When alternate habitat quality is high or the relative risk of predation is particularly high, foragers should give up foraging after depleting the seed trays modestly, whereas foragers should have lower giving-up seed densities when alternative habitats are less attractive (J. S. Brown 1988). A series of such experiments with a variety of granivore species in a variety of environments have consistently demonstrated the same patterns (Brown and Alkon 1990; Brown et al. 1992*a,b*; Kotler 1992; Brown et al. 1994): Foragers depart risky habitats (trays) sooner than safe habitats and depart habitats sooner when alternate feeding areas offer higher food densities.

In summary, a great deal of experimental evidence indicates that habitat selectivity decreases with reduced resource availability or increased forager density, but increases with increasing predator density, in accordance with the simple marginal fitness gains model we (following numerous others) have developed. Little work to our knowledge has addressed whether foragers are attracted to aggregations of conspecifics to dilute their own risk of predation. Perhaps one of the best studies was conducted by Turchin and Kareiva (1989) on the aphid *Aphis varians*, which is subject to attack by a beetle predator, *Hippodamia convergens*. They showed that although both aphids and their predators preferentially seek out aphid aggregations, the dilution effect was more than sufficient to supersede the effect of predator attraction, with large fitness consequences.

4.3.2 Food Chain Dynamics

We evaluated the effect of facultative habitat distribution by herbivores under the assumption that carnivores also exhibit a facultative numerical response. We assumed differences in habitat cover lead to different rates of effective search by both herbivores and carnivores, with faster rates of search in habitat 1 than in habitat 2 ($a_1 > a_2$ and $A_1 > A_2$). Dynamics of the system are dictated by the following set of equations:

$$\frac{dR_1}{dt} = rR_1\left(1 - \frac{R_1}{K}\right) - \frac{\beta a_1 R_1 N}{1 + h[\beta a_1 R_1 + (1 - \beta) a_2 R_2]} \tag{4.16}$$

$$\frac{dR_2}{dt} = rR_2\left(1 - \frac{R_2}{K}\right) - \frac{(1 - \beta) a_2 R_2 N}{1 + h[\beta a_1 R_1 + (1 - \beta) a_2 R_2]} \tag{4.17}$$

$$\frac{dN}{dt} = \frac{ceN[\beta a_1 R_1 + (1 - \beta) a_2 R_2]}{1 + h[\beta a_1 R_1 + (1 - \beta) a_2 R_2]} - \frac{PN[A_1 \gamma \beta + A_2 (1 - \gamma)(1 - \beta)]}{1 + HN[A_1 \gamma \beta + A_2 (1 - \gamma)(1 - \beta)]} - dN \tag{4.18}$$

$$\frac{dP}{dt} = P\left(\frac{CEN[\beta \gamma A_1 + (1 - \beta)(1 - \gamma) A_2]}{1 + HN[\beta \gamma A_1 + (1 - \beta)(1 - \gamma) A_2]} - D\right) \tag{4.19}$$

The optimal habitat selection model for herbivores involves decreased use of the protective dense habitat when resource density in that habitat falls, carnivore density gets too high, or herbivore density falls too low. If carnivores did not select habitats to maximize fitness, then the qualitative conditions favored by natural selection would tend to be stabilizing (Sih 1987). On the other hand, facultative habitat selection by both carnivores and herbivores neutralizes any stabilizing effect: A system without facultative habitat selection (β and γ always set to 0.5) has similar levels of population variability as a system in which both carnivores and herbivores exercise facultative habitat selection (Fig. 4.11). The small set of parameter values at which the habitat selection model *is* more stable have an interesting feature: Stability results from out of phase oscillations in each habitat (Fig. 4.12). As carnivores overwhelm herbivores in any given habitat, herbivores emigrate into the alternate habitat. This refuge doesn't last long, because carnivores rapidly switch to the alternate habitat. Because the oscillations are out of phase across habitats, densities of both predators and prey averaged across habitats vary little over time.

This pattern is identical to that arising via migration among coupled populations with unstable logistic cycles (Hastings 1993). In the coupled oscillator model, Hastings found that parameters producing chaos in an unstructured environment often settled into low amplitude cycles or even point stability if the populations were initiated slightly out of phase. It would be interesting to find how similar this analogy extends, particularly given that 3-species interactions are known to produce chaos for realistic parameter combinations (Hastings and Powell 1991; McCann and Yodzis 1994).

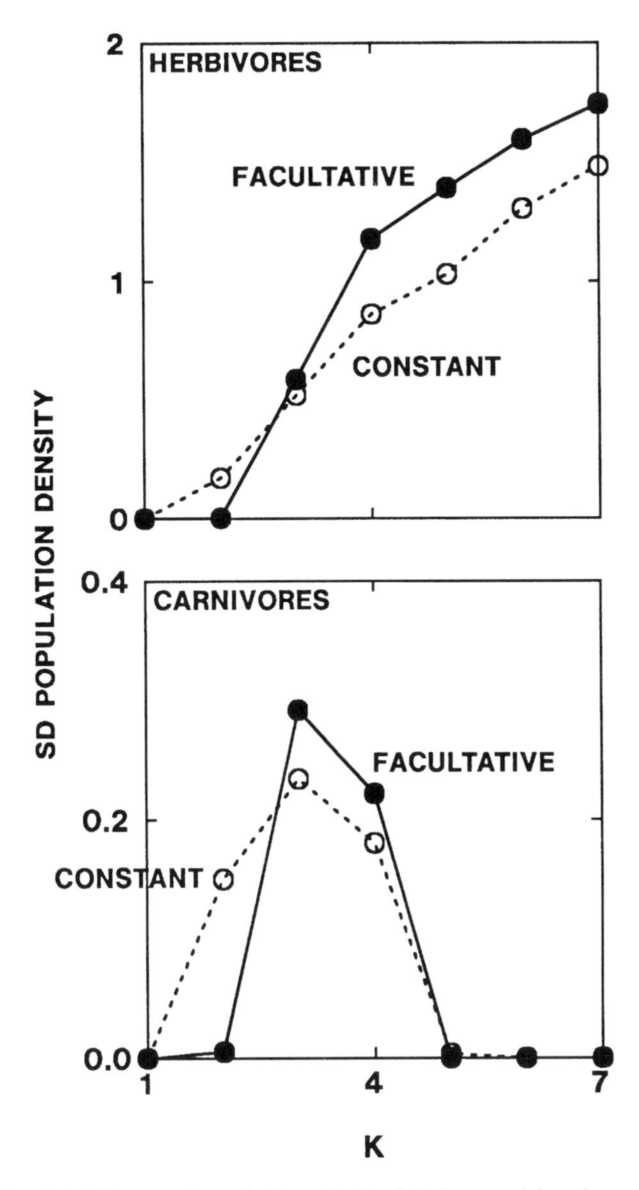

Figure 4.11. Variability over time of a 3-species food chain comprising plants, herbivores, and carnivores in a system with 2 habitats, with facultative habitat choice by both carnivores and herbivores or constant dispersal by carnivores or herbivores. The following parameter values were used: $a_1 = A_1 = 1.2$, $a_2 = A_2 = 0.8$, $z = 4$, $e = E = 2$, $c = C = d = D = r = 1$, $h = H = 0.5$.

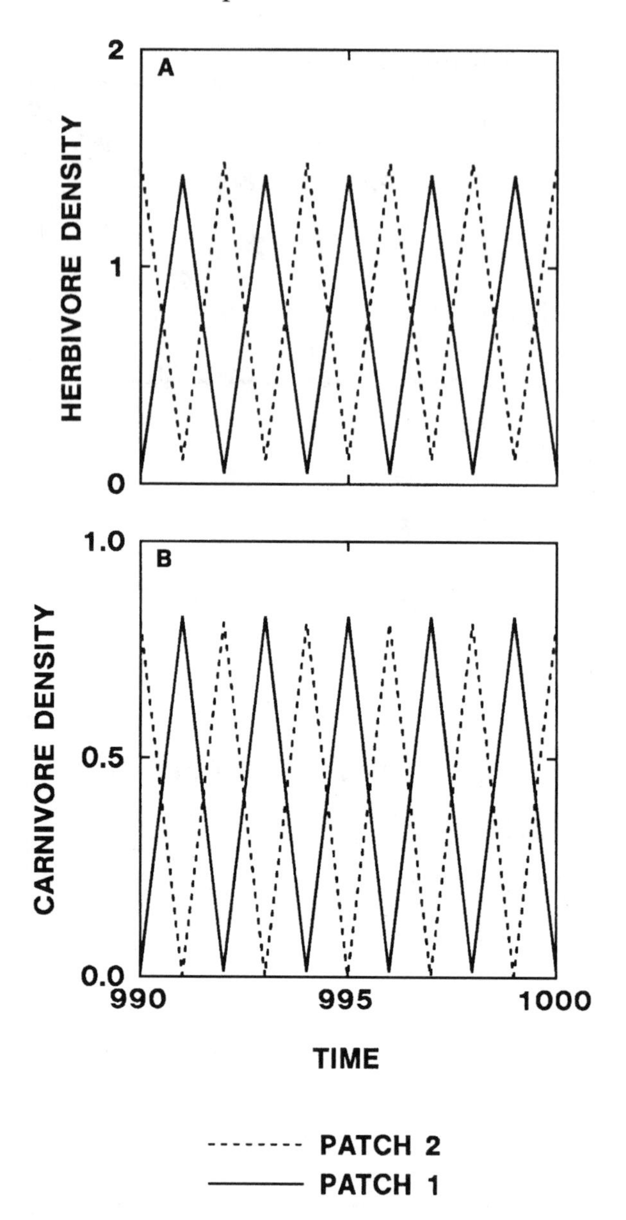

Figure 4.12. Herbivore (A) and carnivore (B) densities within each habitat for a stable 3-species food chain with facultative habitat choice by both carnivores and herbivores. Parameter values as in Fig. 4.11, except $K = 2$.

4.4 SPATIAL STRUCTURE

To this point, we have assumed that habitats vary inherently with respect to rates of resource acquisition and predation risk, while ignoring spatial structure. We now approach the flip side of the issue and assume that all patches are more or less equally suitable as habitats and explore the importance of the spatial structure of suitable patches on trophic interactions. Assume that resource profitability (e/h) and ecological conditions (a, r, and K) are identical in each patch in the environment, but individuals must spend time moving from one patch to the next. If foragers know the mean rate of energy obtainable in other patches in the environment (a debatable assumption in itself), then an energy-maximizing forager should depart a given patch when the rate of energy gain within patch ij equals the expected rate of energy gain (averaged over all other patches in the environment) discounted by the time spent in travel (Charnov 1976b; Rosenzweig 1981; Bernstein et al. 1988, 1991; Green 1990; Fryxell and Lundberg 1993):

$$\frac{eaR}{1 + ahR} = \frac{ea\bar{R}}{(1 + \tau)\,(1 + ah\bar{R})} \tag{4.20}$$

where τ = the time it takes to travel between patches. According to this energy-maximizing rule, foragers should never emigrate unless resource density within their current patch falls well below that in neighboring patches (Fig. 4.13). The magnitude of acceptable deviation between resource density in the current and alternate patches of course depends directly on the "cost" of travel to a new patch. Highly isolated patches would be abandoned at a much lower density of resources than closely packed patches (Charnov 1976b).

4.4.1 Experimental Evidence of Optimal Patch Use

Several nicely controlled studies have been conducted to test the effect of travel costs on patch depression, usually in patches with nonrenewing resources. The classic study is perhaps Cowie's (1977) lab studies of patch use by great tits (*Parus major*). The birds had been taught to forage among sawdust-filled containers hung from the branches of artificial trees in an aviary. When trees were widely spaced, birds spent considerably longer searching within each patch than when trees were closely spaced. There is strong reason to suspect that this pattern reflected diminishing returns within patches, as the instantaneous rate of energy gain declined over time in negative exponential fashion. Hence, birds left isolated patches at a lower rate of gain and lower prey density than closely packed patches.

Numerous studies have shown the predators concentrate their search activity in patches of higher-than-average rates of gain, at least in experimental systems (Milinski 1979; Harper 1982; Recer et al. 1987; Sutherland et al. 1988; Hanson and Green 1989; Korona 1990; Ward and Saltz 1994; Gray and Kennedy 1994; Cuddington and McCauley 1994; Focardi et al. 1996). Travel costs and back-

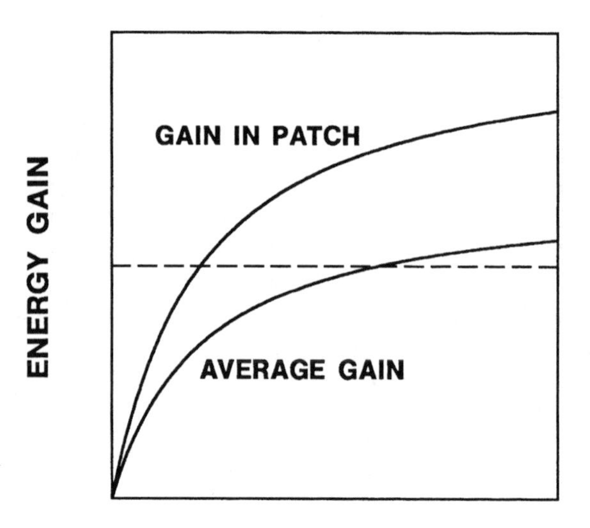

Figure 4.13. Rates of energy gain by a forager remaining in its patch or moving to a neighboring patch in the environment. If resource density in a given patch matches that in surrounding patches, it would never pay to move, because of the time cost of emigration. Provided resource density in a patch falls below that available elsewhere in the environment, it may pay a forager to pay the cost of emigration.

ground resource availability affect the degree of preference and of depletion of high yield patches (Hanson and Green 1989; Korona 1990; Brown et al. 1992*a*).

Numerous studies also suggest that foragers are incapable of perfectly identifying patch quality, with good patches used less frequently and poor patches used more frequently than would be expected on the basis of energy-maximizing principles (Sutherland et al. 1988; Gray and Kennedy 1994). This could stem from lack of perfect information, competitive differences among predators (Parker and Sutherland 1986; Sutherland and Parker 1992), or simply limits on perceptual ability of foragers (Abrahams 1986; Gray and Kennedy 1994). One can phenomenologically model this sloppiness around the perfect energy-maximizing decision by use of the β function (eqn. 4.4). There is reason to believe that sloppy patch use may actually improve fitness in situations in which foragers have imperfect information (Fryxell 1997).

4.4.2 Patchy Predatory–Prey Dynamics

Unlike the previous models we have examined, time must be treated explicitly when modeling a patchy environment, because time must be spent traveling from one site to the next. As well, one must keep explicit track of spatial location, because if we presume it takes a given amount of time to emigrate, then it makes no sense to presume that emigres can travel an infinite distance within that allotted time. Hence, it is more realistic to model movements as a series of short jumps between neighboring patches (the stepping-stone model of movements) rather than global jumps clear across the environment (the island model of movements). In this sense, our approach is more reminiscent of the family of random walk models (Skellam 1951; Okubo 1980; Kareiva and Odell 1987; Bernstein et al. 1988; Turchin 1991; Holmes 1993; Hassell et al. 1991, 1994) than metapopulation models with global movements (Levins and Culver 1971; Hanski and Gilpin 1991; Caswell and Etter 1993; Hanski 1994; Tilman et al. 1994).

We start by considering the spatial dynamics of plants and herbivores. Later in the chapter we add mobile carnivores to the community. We assume local density dependence in the plant population and a monotonically saturating functional response by herbivores, modeling local dynamics in each patch by the following set of equations:

$$\frac{dR_{ij}}{dt} = rR_{ij}\left(1 - \frac{R_{ij}}{K}\right) - \frac{aR_{ij}\beta_{ij}N_{ij}}{1 + ahR_{ij}} \tag{4.21}$$

$$\frac{dN_{ij}}{dt} = \beta_{ij}N_{ij}\left(\frac{ceaR_{ij}}{1 + ahR_{ij}} - d\right) - (1 - \beta_{ij})N_{ij} + I_{ij} \tag{4.22}$$

where R_{ij}, N_{ij}, and I_{ij} are local plant, resident herbivore, and immigrant herbivore densities in the patch with i and j Cartesian coordinates and β_{ij} is the probability that an individual herbivore remains in habitat ij calculated on the basis of the sigmoid function:

$$\beta_{ij} = \frac{e^{z\phi_{ij}}}{e^{z\phi_{ij}} + e^{z\bar{\phi}/(1 + \tau)}} \tag{4.23}$$

where ϕ_{ij} = the rate of energy gain in patch ij and mean ϕ = average energy gain available in the entire environment. We assume that immigrants from neighboring patches lose τ time steps during which their density is subject to the density-dependent survival rate $(1 - d)$ and that emigrants are evenly partitioned among the nearest neighboring patches (assuming no diagonal movements). Hence the density of immigrants entering interior patch ij at time t is calculated by $I_{ij,t} = 0.25\ \Sigma\ (1 - \beta)(1 - d)^{\tau}\mathrm{N}_{i \pm 1, j \pm 1, t - \tau^*}$. We assumed an environment with reflecting boundaries, presuming that emigrants reaching the edge of their suitable range must backtrack rather than move over an infinite plain. This assumption implies that emigrants from edge patches are spread among the 3 neighboring patches and emigrants from corner patches are split between the 2 neighboring patches. Simulations were initiated with plants at carrying capacity in all cells in the grid and a small population of herbivores were introduced simultaneously into a corner patch. We compared outcomes of models with no emigration ($\beta_{ij} = 0$), constant probability of emigration ($\beta_{ij} = 0.05$), or facultative emigration (β_{ij} calculated according to eqn. 4.23).

A constant probability of movement to neighboring patches has little effect, whereas facultative dispersal can have a strong stabilizing influence on plant–herbivore dynamics (Fig. 4.14). In a system with constant dispersal, herbivore increase within a given patch is followed by a gradual spread outward to neighboring patches, with the rate of spread directly a function of the magnitude of the dispersal rate ($1 - \beta$). If local carrying capacities are such that an isolated cell would never approach a stable equilibrium, then the number of emigrants to neighboring cells cycles is in unison with local density. As a consequence, movements tend to synchronize dynamics in neighboring cells. On the other hand, it takes time for emigrants to diffuse across the patchwork, which tends to reduce the correlation between widely separated patches. The outcome at the metapopulation level largely depends on the resulting spatial autocorrelation among cells. When the correlation between cells declines sharply with distance, metapopulation variability is reduced, because cells fluctuate out of phase to some degree. Hence, the aggregate can fluctuate less than that of each of its parts. By and large, models with a constant probability of emigration tend to become synchronized over space and therefore nearly as variable as homogeneous systems.

In a system with facultative emigration, herbivores stay put until they have depressed resources considerably below levels seen in neighboring cells. As a consequence, little dispersal occurs until herbivores have built up to relatively high density, then wholesale emigration occurs to neighboring patches and the process begins anew. This implies that herbivore dynamics are less likely to become synchronized among neighboring patches, because peak predator densities in one patch do not immediately translate into high densities in neighboring areas until N falls well below the average resource density elsewhere (Fryxell and Lundberg 1993). This tends to produce erratic population fluctuations within cells and less spatial synchrony across cells (Fryxell and Lundberg 1993). Because more cells at any given moment are fluctuating out of phase with one another, the metapopulation is often less variable than that of a system with constant probability of dispersal (Fig. 4.14).

It is more likely that synchronization across cells produces metapopulation oscillations when the number of patches is modest or transit time during emigration is small (Fryxell and Lundberg 1993). Cyclical metapopulations often exhibit standing waves of abundance rolling across the landscape (Fig. 4.15), whereas in stable systems, local populations usually approach a heterogeneous, but fixed lattice structure (Hassell et al. 1991; Comins et al. 1992; Hassell et al. 1994).

Metapopulation variability is inversely related to travel costs in systems with facultative emigration (Fryxell and Lundberg 1993), because dispersal reduces the rate of herbivore recruitment at the same time that it reduces the rate of

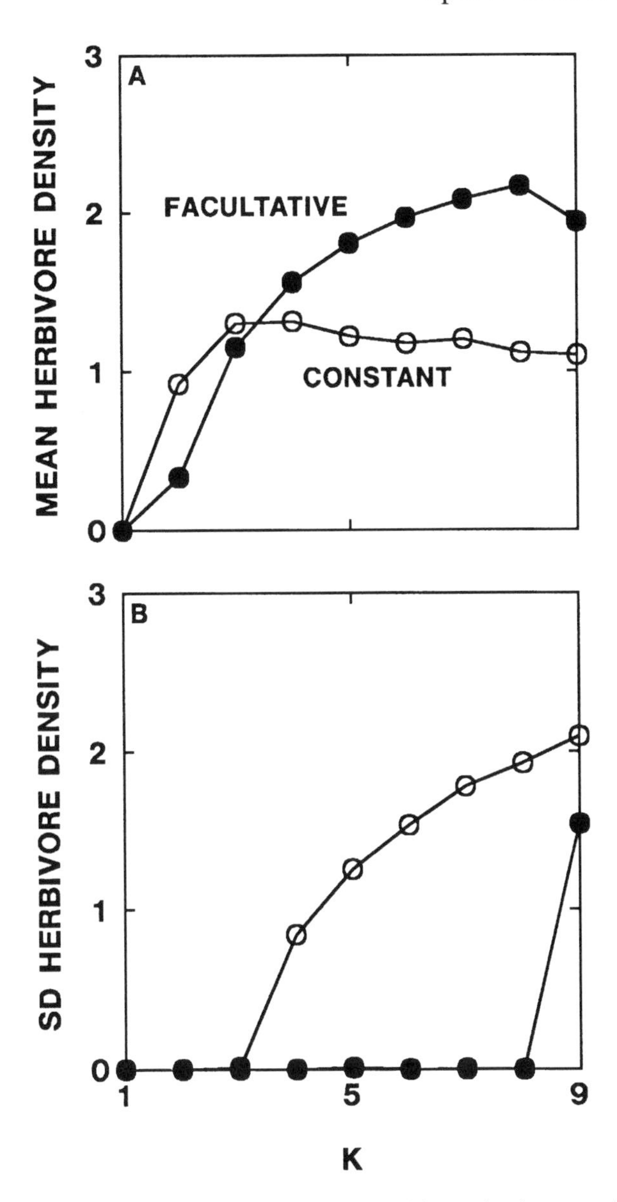

Figure 4.14. Mean herbivore density and SD of herbivore density over time for a 2-species food chain. The SD of herbivore density was recorded over 500 time steps for a metapopulation of constantly dispersing herbivores (open circles), or facultatively dispersing herbivores (closed circles), using a matrix of 49 identical patches. The following parameter values were used: $e = 2$, $a = c = d = r = 1$, $h = 0.5$, $z = 4$, $\tau = 0.08$.

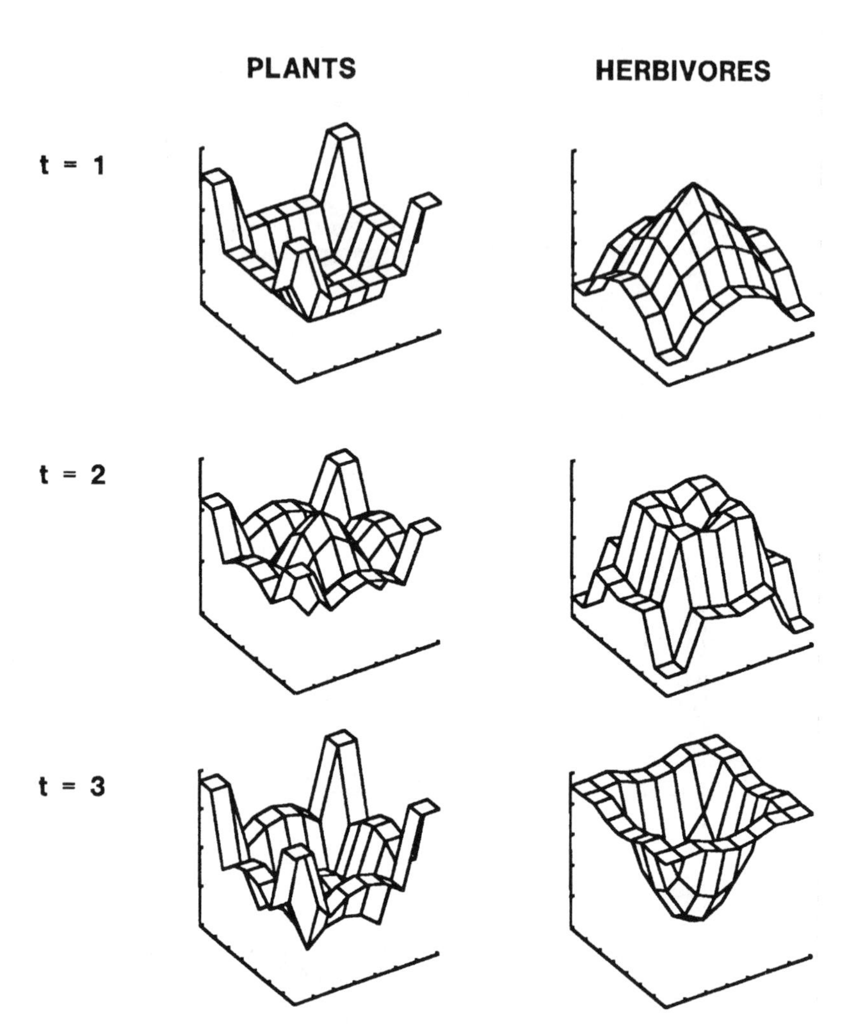

Figure 4.15. Local plant and herbivore densities within each patch at 3 time steps for an unstable system with facultative herbivore dispersal. The following parameter values were used: $K = 8$, $e = 2$, $a = c = d = r = h = 1$, $z = 4$, $\tau = 0.04$.

resource consumption, even though it is advantageous to individual dispersers. The threshold resource density triggering dispersal deviates farther from the environmental average when travel costs are pronounced, which tends to reduce synchrony among patches and decrease metapopulation variability.

We have similarly modeled metapopulations of plants, herbivores, and carnivores in a grid of uniform habitat patches, similar to Schwinning and Rosenzweig (1990) by shifting herbivores or carnivores to neighboring patches whenever fitness within patches falls below the average fitness in all patches in the environment discounted by the dispersal cost τ:

$$\frac{dR_{ij}}{dt} = rR_{ij}\left(1 - \frac{R_{ij}}{K}\right) - \frac{aR_{ij}\beta_{ij}N_{ij}}{1 + ahR_{ij}} \tag{4.24}$$

$$\frac{dN_{ij}}{dt} = \beta_{ij}N_{ij}\left(\frac{ceaR_{ij}}{1 + ahR_{ij}} - d\right) + I_{ij} - \frac{A\beta_{ij}N_{ij}\gamma_{ij}P}{1 + AH\beta_{ij}N_{ij}} - (1 - \beta_{ij})N_{ij} \tag{4.25}$$

$$\frac{dP_{ij}}{dt} = \gamma_{ij}P_{ij}\left(\frac{CEA\beta_{ij}N_{ij}}{1 + AH\beta_{ij}N_{ij}} - D\right) + Y_{ij} - (1 - \gamma_{ij})P_{ij} \tag{4.26}$$

where R_{ij}, N_{ij}, P_{ij}, are local densities of plants, herbivores, and carnivores in patch ij and both β and γ are functions of fitness in patch ij relative to fitness elsewhere discounted by travel time.

Our results suggest that metapopulation dynamics of the 3-species chain with facultative dispersal by both carnivores and herbivores have different patterns of variation than systems with constant dispersal (Fig. 4.16). Facultatively pursuing carnivores have wider variation in abundance than carnivores with constant probability of dispersal. Facultatively evasive herbivores are less variable than herbivores with constant probability of dispersal. Spatial structure also enhances the persistence of carnivores that would otherwise be driven extinct (Fig. 4.17). As in the case of 2-species food chains, 3-species chains tend to be more stable when travel costs are moderately long. In a system with constant dispersal, carnivores risk extinction because their density fluctuates so widely over time. Short periods of positive population growth are simply inadequate to compensate for episodes of low herbivore density. Hence, cycles cascade from plants to herbivores to carnivores, with the latter unable to persist when the system fluctuates too wildly. In the spatially structured system, however, heterogeneous herbivore distributions ensure that there are always enough prey to allow carnivore persistence in some patches, if not all. Interestingly, herbivores tend to be much more mobile than carnivores (Fig. 4.18), which can be traced to the fact that herbivore fitness is shaped by both food availability and their risk of being eaten by carnivores. As in the case of 2-chain models, unstable metapopulations usually exhibit population waves rolling periodically across the landscape (Fig. 4.18).

One reason that complete metapopulation stabilization is rarely achieved is that the "ideal-free distribution" is unlikely in even 2-species food chains with explicit spatial structure, let alone longer food chains (Schwinning and Rosenzweig 1990; Bernstein et al. 1991). It is not too surprising, therefore, that stability is not a particularly likely outcome.

It is somewhat difficult to compare our results with previous predator–prey models incorporating spatial structure. There are several elegant models of pure

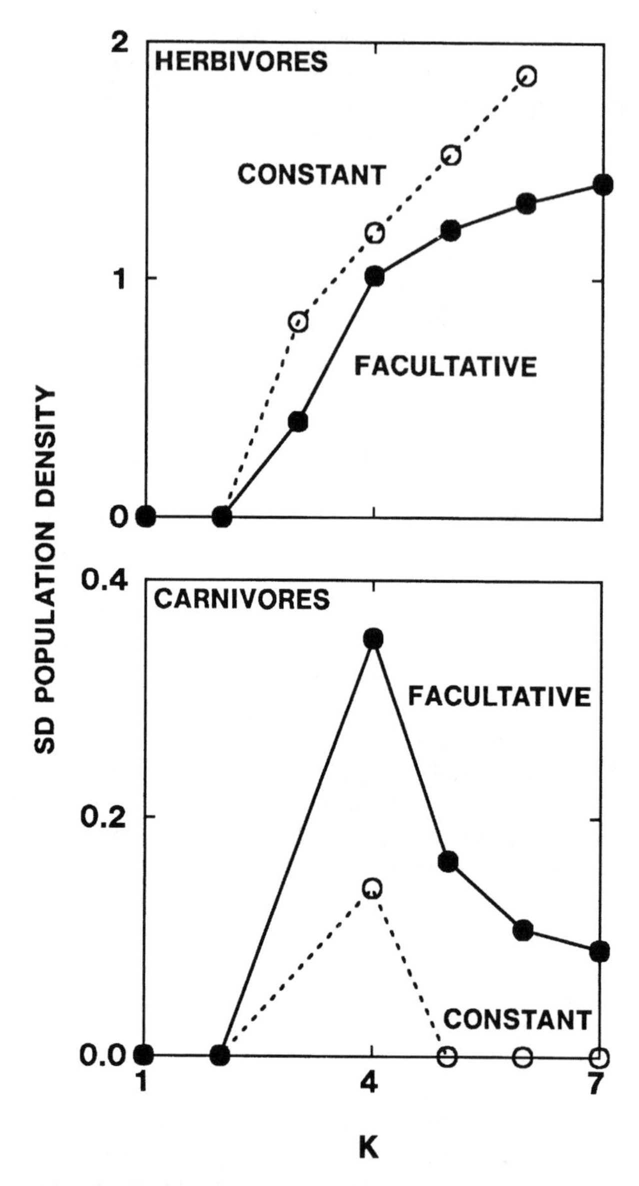

Figure 4.16. Variability over time for a 3-species food chain comprising plants, herbivores, and carnivores, with both carnivores and herbivores with either facultative dispersal (closed circles) or constant dispersal (at 5% per time step). SD of herbivore and carnivore densities was recorded over 500 time steps for a metapopulation in a matrix of 49 identical patches. The following parameter values were used: $a_1 = A_1 = 1.2$, $a_2 = A_2 = 0.8$, $c = C = d = D = r = 1$, $e = E = 2$, $h = H = 0.5$, $z = 4$, $\tau = 0.04$.

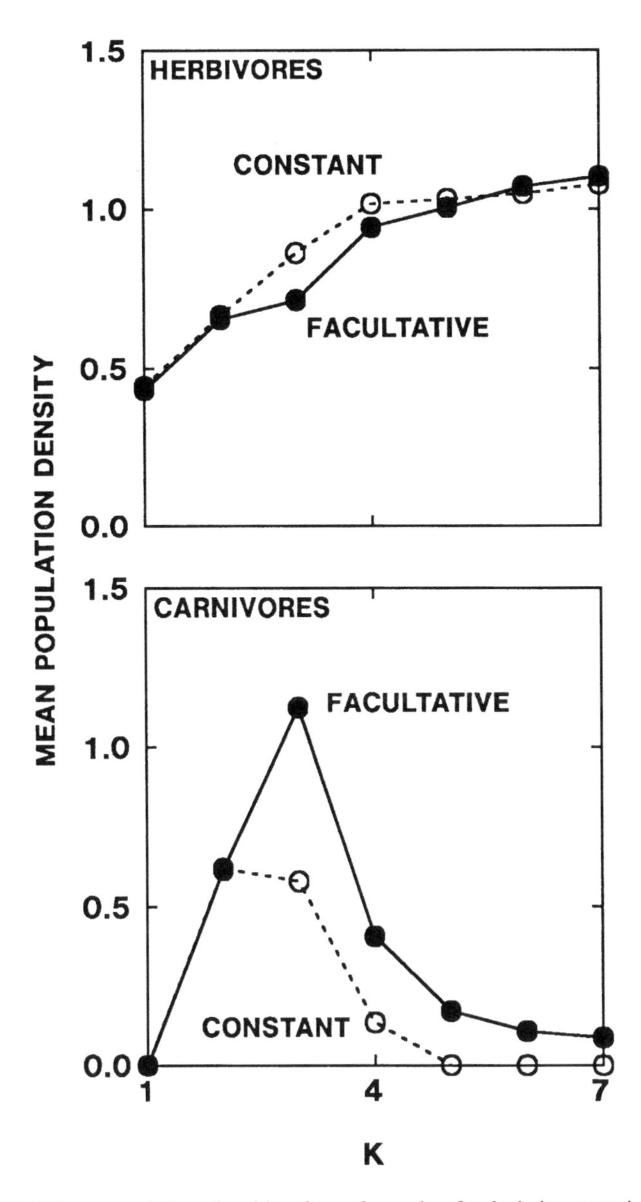

Figure 4.17. Mean population densities for a 3-species food chain comprising plants, herbivores, and carnivores, with both carnivores and herbivores with either facultative dispersal (closed circles) or constant dispersal (at 5% per time step). Mean herbivore and carnivore densities were recorded over 500 time steps for a metapopulation in a matrix of 49 identical patches. Parameters were the same as in Fig. 4.16.

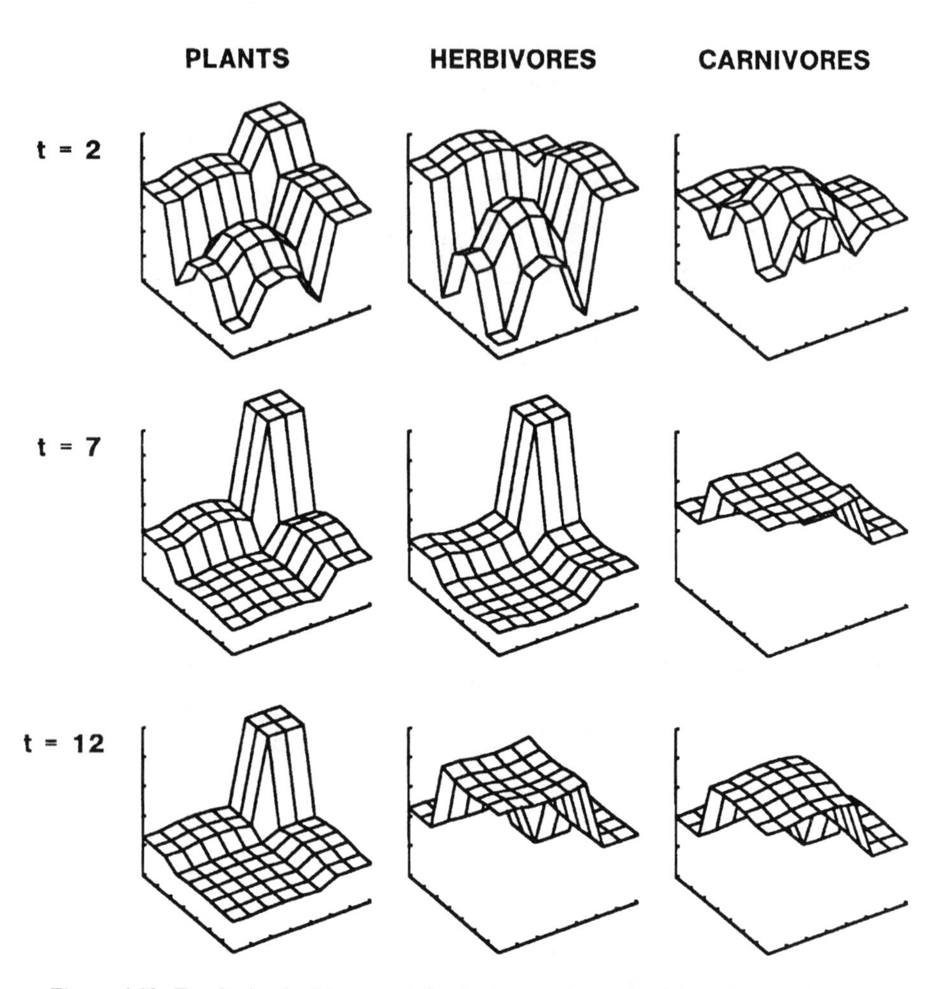

Figure 4.18. Facultative herbivore and facultative carnivore densities within each patch at three time steps for an unstable 3-species food chain inhabiting a metapopulation matrix of 49 identical patches. Parameters were the same as in Fig. 4.16 except $K = 8$.

diffusion by predators, prey, or both (Comins and Blatt 1974; Levin 1974; Mimura and Murray 1978; Okubo 1980; McLaughlin and Roughgarden 1991) or approximations using individual-based lattice simulation models (de Roos et al. 1991; McCauley et al. 1993). Under some conditions, such models can generate heterogeneity in predator and prey spatial distributions in an otherwise continuous and constant environment, analogous to the asynchronous dynamics arising in the metapopulation model with facultative dispersal. Diffusion models are difficult to solve in unstable systems, but the lattice simulation analog strongly suggests that pronounced spatial asynchrony is associated with reduced population variability, as we find in the metapopulation model (de Roos et al. 1991; McCauley et al. 1993).

Another large class of spatially structured models relates to systems with equal probabilities of dispersal to any patch (Levins and Culver 1991; Ziegler 1977; Hastings 1977; Gurney and Nisbet 1978; Crowley 1981; Hogeweg and Hesper 1981; Hanski and Gilpin 1991; Caswell and Etter 1993; Nisbet et al. 1993; Hanski 1994; Tilman et al. 1994). Such classic metapopulation models are conceptually more similar to our models of two coupled habitats than models with true spatial structure, because immigrants can arise anywhere outside a focal patch. Such models do suggest that the size of the patch ensemble and asymmetry in dispersal rates of predators and prey are critical in preventing metapopulation extinction, in agreement with our patch models.

Perhaps the closest models in spirit are recent models of parasitoid and host populations with local dispersal (Hassell et al. 1991; Comins et al. 1992; Hassell et al. 1994) and insect predator–prey models (Hilborn 1975; Nachman 1987; Sabelis et al. 1991). This work suggests that spatially structured populations of parasitoids or predators may be more persistent, even if they are not particularly stable per se. The parasitoid–host models also suggest that metapopulations have the potential to develop complex spatial formations that simply don't arise without localized dispersal processes and hint that such spatial structures strongly influence metapopulation variability, via asynchrony of distant patches.

Regardless of model structure, there seems to be nearly universal acceptance that the degree of synchrony among spatial locations is an essential determinant of metapopulation dynamics. We differ, however, from most others in emphasizing the dynamical importance of facultative emigration strategies. Nearly all previous metapopulation models, regardless of structure, incorporate constant probabilities of emigration. The assumption of constant dispersal would be maladaptive under most circumstances (Holt 1983, 1985; van Baalen and Sabelis 1993), suggesting that facultative emigration may be slightly more plausible. A troublesome problem lurking behind all metapopulation studies is the possibility of long-lived transient cyclical or even chaotic behavior in even single-species models with complex spatial structure (Hastings and Higgins 1994).

4.4.3 Experimental Evidence of Patchy Predator–Prey Dynamics

By and large, theoretical understanding of the potential dynamics of spatially structured systems vastly outstrips the empirical evidence (Kareiva 1990; Taylor 1991). Indeed, there is a real danger that metapopulation theory becomes so firmly entrenched in ecological dogma that biologists simply take it on faith (Kareiva and Wennergren 1995). The existing evidence is both fragmentary and contradictory.

One of the best and well-known empirical studies of spatial dynamics is Huffaker's (1958) classic experiments with herbivorous and carnivorous mites inhabiting a universe of oranges (i.e., patches) in the laboratory. Through a painstaking series of microcosm experiments, Huffaker found that persistence time was enhanced by complicated patch arrangements, particularly when dispersal by predatory mites was discouraged by the use of barriers between oranges. Mite populations ultimately went extinct in all of Huffaker's experiments. Nonetheless, his conclusion that spatial complexity encourages longer persistence is at least qualitatively consistent with the facultative dispersal models.

More recent experiments with greenhouse populations of herbivorous and carnivorous mites have extended Huffaker's results to more realistic environmental systems (Nachman 1991). Isolated plants rarely support both herbivorous and carnivorous mites for more than a few weeks, exhibiting a classic herbivore explosion followed by abrupt crash of both herbivore and carnivore, whereas whole greenhouse populations persist, albeit unstably, for several months. Observations at frequent intervals throughout the greenhouses indicated asynchronous dynamics between regions, with spatial autocorrelation declining with distance between patches (Nachman 1991). It may seem inconceivable that regional dynamics would differ across the width of a greenhouse, but the average area visited during the lifespan of a single carnivorous mite is minuscule (ca. 5m^2). Associated behavioral experiments showed that dispersal rates in both the predatory and herbivorous mites were negatively related to prey density (Bernstein 1984). Interestingly, dispersal rates by the herbivorous mites were also sensitive to densities of the carnivorous mites, consistent with our facultative behavioral response by a forager trapped between the "devil and the deep blue sea." The greenhouse system of cucumber plants, herbivorous mites, and carnivorous mites therefore has all the requisite characteristics of a shifting mosaic metapopulation. It is intriguing that their spatially structured experimental system, like most model systems, is more persistent than its homogeneous counterpart, but not particularly stable.

There are also well-studied examples of patchy systems whose dynamics are inconsistent with our metapopulation model (Taylor 1991). One of the most elegant experiments was conducted by Kareiva (1987) on a 3-species food chain consisting of goldenrod plants, herbivorous aphids, and carnivorous beetles. Detailed observations of predatory search behavior indicated that beetles altered their turning frequency and search velocity in a manner leading to aggregation in the vicinity of local outbreaks of aphids (Kareiva and Odell 1987), a pattern verified in a later field experiment (Vail 1993). This "preytaxis" behavior is analogous to the facultative dispersal behavior in our metapopulation model. Over the course of three summers, Kareiva compared the dynamics of continuous strips of goldenrod versus patchy strips of goldenrod, with results showing that aphid and beetle densities were highest in the fragmented landscape. Inoculated plants in both treatments showed that fragmentation led more frequently to aphid outbreaks and less frequently to biological control by the beetles (Kareiva 1987).

Hence, in the goldenrod–aphid–beetle system, patchiness was associated with increased variability rather than reduced variability (Kareiva 1990), in contrast to our metapopulation model. Apparently patch separation that is too extreme accentuates, rather than reduces, population variability, perhaps because beetles fail to find neighboring patches or beetles rarely disperse at all. In any case, Kareiva's work and the more recent studies of Vail (1993) on essentially the same system clearly demonstrate the need to consider the spatial scale over which resources vary relative to the spatial scale of the predator's behavioral response.

4.5 SUMMARY

Ecologists have long recognized that environments are usually highly heterogeneous, but relatively little attention has been focused on the dynamical implications of energy-maximizing behavioral responses to that variability. We review decision rules for optimal habitat use to maximize the rate of energy gain. We then extend those decision rules to include mortality risk.

Optimal habitat use can have a stabilizing influence on community dynamics only if the habitats have different dynamics in isolation. Suppose one habitat in isolation would support stable populations, but the other habitat in isolation produces cycling populations of plants and herbivores. A coupled system of one stable and one unstable habitat can be stabilized by facultative emigration by herbivores from one habitat to the next. This stabilizing influence is not nearly as strong, however, when carnivores are added to the system that facultatively shift habitats to maximize fitness.

We extended this basic model to consider mechanistic competition among consumers with different habitat use strategies. The ESS community changed with ecological parameters, sometimes favoring specialists in the better habitat, sometimes favoring a mix of opportunistic, facultative, and specialist strategies, and sometimes only selecting for facultative habitat choice. Stabilization of communities due to facultative habitat shifts allowed the widest variety of competitors to persist in the system, due to the fact that many strategies would have equal fitness at equilibrium—the fitness landscape is essentially flat.

We also consider the spatial dynamics of a metapopulation of patches with local populations of plants, herbivores, and carnivores, linked via facultative emigration by herbivores and carnivores. Decision rules for facultative dispersal were developed from the marginal value theorem, comparing the fitness within patches with the expected fitness elsewhere in the environment scaled by travel costs. Simple systems with plants interacting with mobile herbivores could be stabilized substantially by facultative emigration rules. When mobile carnivores are added to the system, much of this stabilizing effect largely disappears. Facultative movements among patches do enhance the long-term persistence of herbivores and carnivores in the 3-link system.

5 Size-Selective Predation

All multicellular organisms grow in size during their lifetime. The relationship between time, age, and size and the biological attributes associated with the various stages in life is one of the fundamental issues in life history theory (Stearns 1992; Roff 1994). For example, many plants and animals cannot reproduce unless they have attained a minimum body size. Moreover, the rate of mortality is often strongly associated with body size.

It is obvious that size matters considerably for the dynamics of populations; size and demography are inevitably intertwined. Size variation within most species is considerable, even after accounting for age effects. The relative size difference between "small" and "large" individuals, once they have passed juvenile stages of development, can span several orders of magnitude in some taxa, such as trees. In other organisms, such as birds and small mammals, this difference is minuscule. Demographic life-tables are commonly constructed from age-specific schedules of birth and mortality, which is reasonable if age and size covary. The fate of an individual is often more strongly correlated with its size, however, than its age (e.g., Lefkovitch 1965; Connell 1973; Werner 1975; Harper 1977; Werner and Caswell 1977; Hughes 1984; Caswell 1989). Widespread variation in the interrelationships among life history traits in relation to environmental constraints has lead to a rather substantial literature on the ecology of size-structured populations (e.g., Polis 1984; Werner 1986; Metz and Diekmann 1986; Stearns and Koella 1986; Ebenman and Persson 1988; DeAngelis and Gross 1992; Abrams 1994*b*).

Size structure complicates the analysis of population processes, even when only a single population is considered. The intractability of stage- or age-structured populations has encouraged the development of new families of models. Individual-based models in which each and every individual in the population is kept track of and its fate is specified probabilistically, have become particularly popular in recent years among population ecologists (DeAngelis and Gross 1992). Coupled consumer-resource models of size-structured populations are less common. This is perhaps not surprising. Introducing size structure into trophic models of community dynamics requires critical assumptions about mechanisms of intraspecific competition (e.g., density dependence) as well as detailed assumptions about the foraging process of the consumer. In some sense each size class functions as a separate species, hence even simple models of size-structured populations of predators and prey have the complexity of a food web. This increase in dimensionality inevitably makes the analysis of size-structured systems cumbersome (Caswell 1989). Moreover, it is often challenging to model the internal dynamics of size-structured populations. Density-dependent relationships may be particularly complex because of size-structured patterns of intraspecific competition.

In this chapter, we address the dynamics of some simple models of size-selective predation. We envision the following general scenario. The resource population, be it plants or animals, can be divided into size classes. Each size class is associated with a given probability of giving birth to new individuals, as well as a given risk of being attacked by predators. There is also a finite probability that an individual of a certain size class moves into another size class. Transition from one size class to another is contingent on a number of factors that we incorporate into full dynamical models of population interactions. For simplicity, we assume that only the resource population is size structured and that any member of the consumer population is at least potentially capable of attacking any member of the resource population. That is, the consumer population is assumed to have no size structure. Otherwise, our models would become very cumbersome.

We will consider 2 size-selective models. In the first scenario, which we call the diet selection model, the forager must decide which size classes to include in its diet by following an energy-maximizing strategy. In the second scenario, which we call the partial predation model, the forager treats each prey item as a patch, which can be exploited for variable amounts of time in order to maximize rates of long-term energy gain.

5.1 DIET SELECTION MODEL

5.1.1 Self-Thinning

Intraspecific competition in a size-structured population often leads to what is commonly called self-thinning (White and Harper 1970; White 1977; Harper 1977; Norberg 1988; Lonsdale 1990). Although self-thinning is generally thought of as being a phenomenon of plants, there is increasing evidence that many, if not most, animal populations also show the same pattern (Latto 1994). This suggests that self-thinning is nothing more than an expression of intraspecific competition (Watkinson 1980). A constant resource supply in a given physical area probably has limits on the maximum sustainable biomass, which we can think of as the carrying capacity (K). This biomass may divided into a few large individuals or many small ones. High densities of individuals are associated with small average size. Conversely, as individuals in the population grow bigger, their density must decline. Such processes would result in the empirically well-confirmed $-3/2$ power law (Taitt 1988).

Self-thinning can be modeled in the following manner. Suppose that the population is divided into I size classes (b_i, where $i = 1, 2, \ldots I$) of density N_i and that the largest class is of size $b_{\max}$. We need to specify how mass changes with each size class. From the wide range of suitable models, we use a simple logistic curve:

$$b_{i+1} = gb_i\left(1 - \frac{b_i}{b_{max}}\right) \quad (5.1)$$

where g = a mass-specific growth parameter. So long as $\Sigma b_i N_i < K$, the density and/or size of individuals can increase. In order to account for this, the rate of change of the number of individuals will then be a slightly modified form of the logistic. The number of new individuals recruited into the smallest size class (i = 1, implying recruitment to the population) can be specified by

$$\frac{dN_1}{dt} = \sum_{i=2}^{I} r\left(\frac{b_i}{b_{max}}\right)N_i\left(1 - \sum_{i=1}^{I}\frac{b_i N_i}{K}\right) \quad (5.2)$$

where r is the maximum per capita rate of reproduction. The ratio b_i/b_{max} implies that large individuals contribute proportionally more to reproduction than small individuals, as well-established in both the plant and animal literature (Charlesworth 1980; Roff 1992; Stearns 1992). Figure 5.1 shows the trajectory over time of mean prey biomass and total prey density for a population growing according to this self-thinning relationship, with size class trajectories illustrated in Figure 5.2.

5.1.2 Size-Dependent Consumption

There is overwhelming evidence for size-selective predation, suggesting that selectivity is associated with variation in energetic profitability (see extensive reviews by Krebs and Davies 1981; Taylor 1984; Begon and Mortimer 1986; Stephens and Krebs 1986; Begon et al. 1996). In a classic study of bluegill sunfish (*Lepomis macrochirus*) feeding on *Daphnia* of three different size classes, Werner and Hall (1974) were able to predict the proportion of large, medium, and small prey in the diet on the basis of relative prey abundance and relative prey size, because in this case size is synonymous with profitability. Krebs et al. (1977) studying great tits (*Parus major*) feeding on mealworms of different sizes also came to the same conclusion: Predators used size as a surrogate measure of profitability and predation resulted from optimal diet selection in relation to prey relative abundance. Elner and Hughes (1978) showed that shore crabs (*Carcinus maenas*) prefer the size of mussel (*Mytilus*) yielding the highest rate of energy gain.

In some circumstances, however, size selectivity can be unreliable. Apparent size varies with distance between the predator and the prey. This source of error can cause trouble for selective planktivorous fish (e.g., Butler and Bence 1984; Wetterer and Bishop 1985).

Before coupling a dynamic predator population with a size-structured prey population, we have to make some additional assumptions. First, let us imagine that the proportion of prey mass available to the consumer varies with prey size.

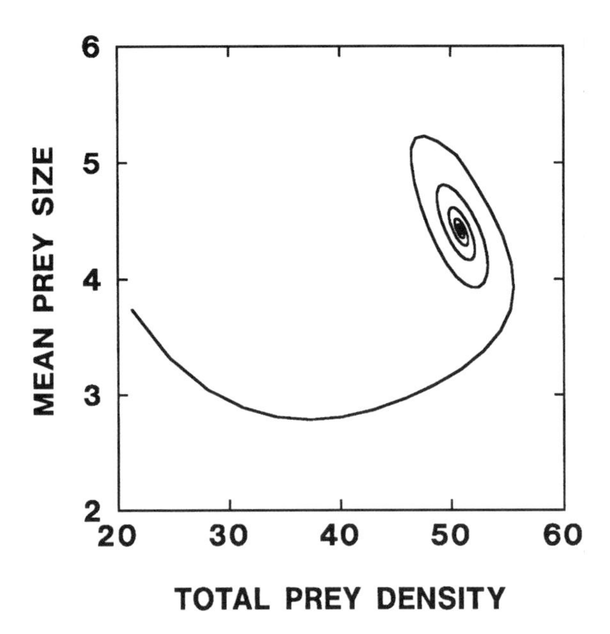

Figure 5.1. The relationship between mean size in the population (with 20 size classes) and the total population density in a resource population with self-thinning and without predator-induced mortality followed over time (500 time steps). Initially, mean size drops and total population density increases succeeded by a concomitant increase in both as the population is established in the habitat. Eventually self-thinning comes into play and there is a trade-off between size and density, finally settling down to an equilibrium. The following parameter values were used: $K = 300$, $b_{max} = 10$, $r = 1.0$.

This proportion could either increase or decrease with size. For many organisms, this proportion would decrease as size increases, because a larger proportion of the body might be made up of support tissue. For instance, much of a plant's mass is tied up in unusable structural tissue, such as stems and branches or the fiber within leaves, hence the edible proportion declines with size. One simple way to model such a structural effect is to calculate the proportion of available tissue p_i of a prey of size b_i by

$$p_i = \frac{1}{(1 + \lambda b_i)} \tag{5.3}$$

where λ is a positive parameter dictating how fast the proportion of available body mass decreases with size. This function has a maximum value of 1 for small prey, implying that the entire prey item is usable by the predator, but decelerates as it approaches an asymptote of 0, implying that all of the tissue is inaccessible to the predator. One can also let profitability vary with size. A large

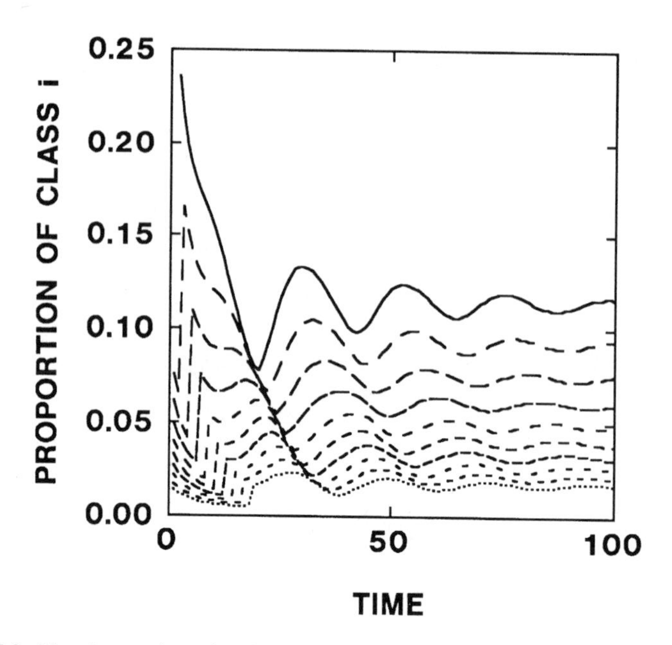

Figure 5.2. The time trajectories for every second size class in a resource population with self-thinning and without mortality due to predation. The population parameters were set so that the population eventually reaches a stable equilibrium. Note that the densities of the different size classes are desynchronized as an effect of self-thinning. The total number of size classes is 20. Parameter values as in Figure 5.1.

prey item obviously provides more food for the consumer than a small one, but profitability would be unchanged if handling time and energetic concentration scale identically with the mass of prey items. It is often the case, however, that energy or nutrient concentration vary with prey mass, particularly in plants (Fritz and Simms 1992 and references therein for review). We therefore define the energy concentration of an individual in size class i (e_i) by

$$e_i = b_i(\gamma + qb_i) \tag{5.4}$$

where γ is the intercept of an arbitrary linear function relating energy concentration to size class and q is the slope of that line. Note that we are using "concentration" in lieu of "content," because e is a measure of energy or nutrients per unit body mass. Energy content and concentration would scale equivalently for prey of a single size, as in traditional predator–prey models. We further assume that the time it takes to handle a single item is linearly related to available mass, on the presumption that prey processing composes the bulk of handling time, a reasonable assumption for herbivores at least (Fryxell et al. 1994). More precisely, handling time per individual of size i (h_i) will be proportional to the available edible size ($b_i p_i$), and we thus write $h_i = \alpha b_i p_i$, where the parameter α scales the handling time per prey.

We are now at the point at which we can calculate the consumption rate. The following expression predicts the number of size class i individuals attacked per consumer:

$$X_i = \frac{\beta_i a N_i}{1 + \alpha a \Sigma \beta_i b_i p_i N_i} \tag{5.5}$$

where a = the area searched per unit time, equivalent for all size classes. This is basically the multispecies disc equation (see Chapter 1) with a few additional features.

β_i is the probability of attack on size class i. If the forager indiscriminately attacks all size classes, then $\beta_i = 1$ for all i. Otherwise, β_i takes on a value depending on the relative profitability of size class i, as it did in our previous models of diet choice among different prey species (Chapter 2). In order to calculate β_i, we first need to rank (r_i) all size classes from 1 for the most profitable size class on through to the least profitable size class. Using the contingency model, we then calculate the long-term gain obtained by sequentially adding prey to the diet in rank order, until we determine the optimal diet breadth, which we will symbolize by ρ. Using the same logic that we used in Chapter 2 to model variation around the optimal diet breadth, we calculate the probability of attack by the following equation:

$$\beta_i = \frac{\exp(z\rho)}{\exp(zr_i) + \exp(z\rho)} \tag{5.6}$$

Under some circumstances, equation 5.5 adequately represents prey mortality due to a single predator. Prey individuals need not necessarily be killed by predator attack, such as when herbivores feed on plants. We therefore let the probability of death following an attack be directly related to the proportion of prey biomass consumed by the forager, such that mortality risk of an attacked individual of size class i is mp_i, where m is a scaling parameter. Since p is a decreasing function of size, big individuals will be less vulnerable to mortality than small individuals. The total number of individuals that die in size class i then becomes $X_i m p_i P$, where P is consumer density.

If an attacked individual does not necessarily die, there is another possible fate. The mass consumed during an attack may be so slight that an individual of size class i remains in that size class. On the other hand, the attack may not be severe enough to kill the individual but sufficiently severe to reduce the prey individual to a smaller size class. That is, prey items could shrink after attack. The number of individuals in size class i that survive attack is $X_i(1 - mp_i)$. The size of an attacked individual changes from b_i to $b_i(1 - p_i)$. If $b_i(1 - p_i)$ falls below one half the size difference between class i and $i-1$, then the surviving $X_i(1-mp_i)$ individuals from class i belong to class $i-1$.

We assume at this point that predators do not choose how much tissue to eat from each prey item. This debatable assumption will be shortly relaxed, when we consider the problem of partial consumption behavior later in this chapter (section 5.2).

This model raises an additional issue of some practical importance. In traditional predator–prey models, the coupling between predation and prey mortality is unambiguous. A prey item that is attacked is inevitably removed from the prey population and converted into new predators. This uncomplicated mapping between predation rate and prey mortality rate falls apart in plant–herbivore systems. Plants are rarely killed after attack (e.g., Hjältén et al. 1993; Bach 1994; Begon et al. 1996), so a different relationship has to be worked out in order to understand dynamic interactions between herbivores and plants. It is against this largely unexplored background that we have presented our size-structured predation model.

5.1.3 Size Selection and Population Dynamics

The rather complicated features of the prey population can now be combined with the familiar equation for the predator population to complete the model community:

$$N_1(t+1) = \sum_{i=2}^{I} r\left(\frac{b_i}{b_{max}}\right) N_i(t)\left(1 - \sum_{i=1}^{I}\frac{b_i p_i}{K}\right) + P(t)\sum_{j=1}^{I} X_j(1-mp_j)S_{1j} - P(t)X_1 mp_1 \tag{5.7}$$

$$N_{i+1}(t+1) = N_i(t) - P(t)X_i mp_i + P(t)\sum_{j=i+1}^{I} X_j(1-mp_j)S_{ij} \quad \textit{for all } i \tag{5.8}$$

$$P(t+1) = P(t)\left(1 + \sum_{i=1}^{I} X_i(t) c e_i b_i p_i - d\right) \tag{5.9}$$

where S_{ij} is the probability (0 or 1) that an individual initially of size class j would shrink to size class i after being eaten. Although this system has a fairly small set of assumptions, it nonetheless accounts for a number of features in a size-structured interaction. Here we concentrate on the effects of selective versus nonselective foraging on community dynamics.

The influence of facultative and nonfacultative foraging on community dynamics is summarized in Figures 5.3 and 5.4. We compare two important aspects of the dynamics. First, we note that there is a small, but consistent, difference between the variability of communities with selective versus nonselective predators, as demonstrated in Figure 5.3 illustrating standard deviation in predator density in relation to prey carrying capacity and the slope of the relationship

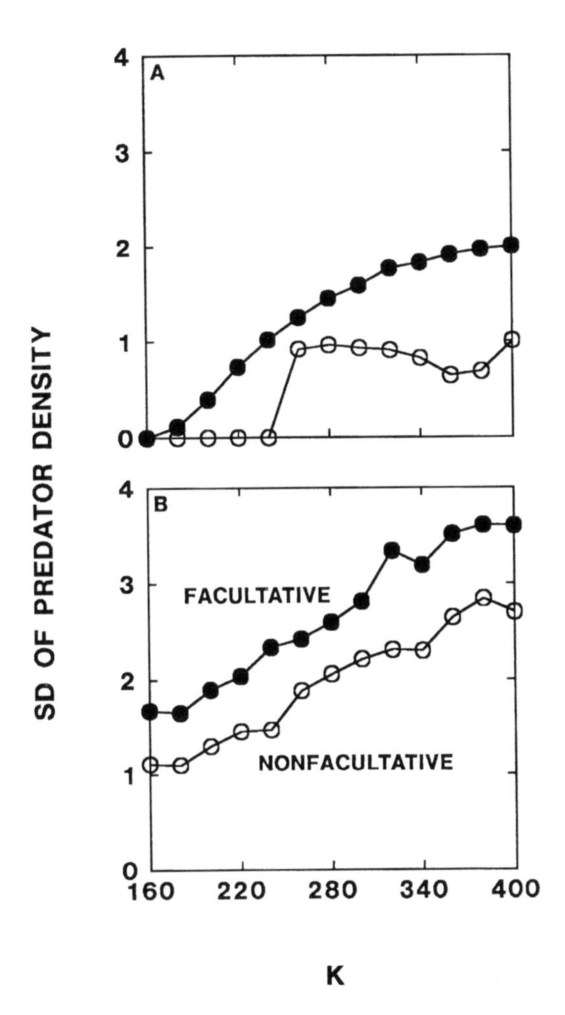

Figure 5.3. Standard deviation of predator densities recorded over the last 500 time steps in 1000 time step simulations as a function of prey carrying capacity K for nonfacultative (open symbols) and facultative (filled symbols) predators. The dynamics of the system is specified by equations 5.7–5.9. Standard deviations are shown for different values of q, the slope of the relationship between energy concentration and size (A, $q = -0.2$; B, $q = 0.2$). Other parameter values were as follows: $K = 300$, $r = 1.0$, $b_{max} = 10$, $a = 0.025$, $\lambda = 0.2$, $d = 0.9$, $c = 0.2$, $\alpha = 0.1$, $m = 0.9$.

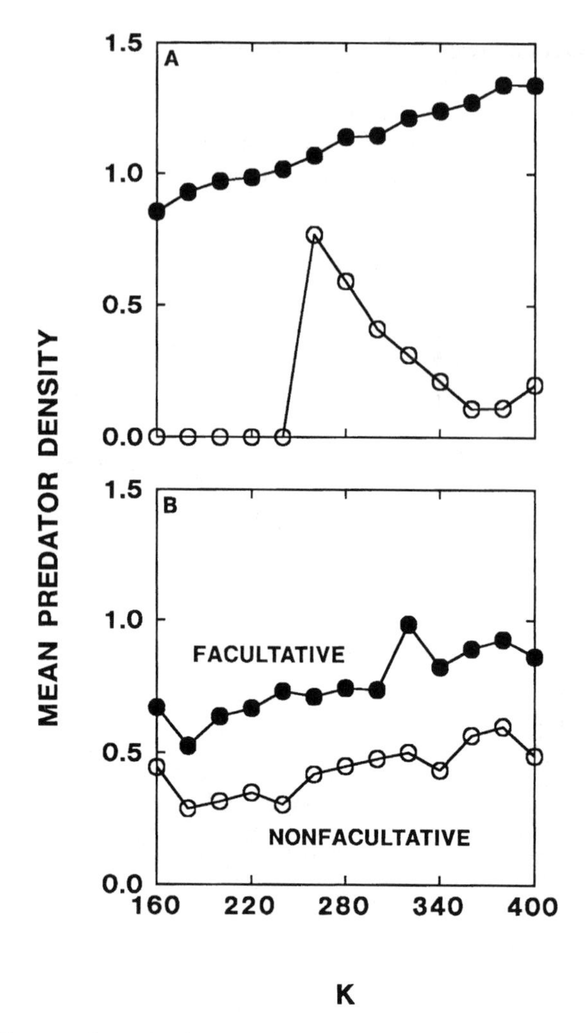

Figure 5.4. Mean predator density recorded over the last 500 time steps in 1000 time step simulations as a function of prey carrying capacity K for nonfacultative (open symbols) and facultative (filled symbols) predators. The dynamics of the system is specified by equations 5.7–5.9. Mean densities are shown for different values of q, the slope of the relationship between energy concentration and size (A, $q = -0.2$; B, $q = 0.2$). Parameter values as in Figure 5.3.

between prey size and energy content. Population variability generally increases with positive slopes. If energetic profitability increases with size, then size-selective predation should increase for large individuals, whereas smaller individuals will be less affected by consumption. Large individuals are more resistant to mortality, because a smaller fraction of their mass is available to foragers.

Selectivity for large prey therefore may tend to boost the prey capacity for growth, promoting greater variability over time.

Although population variability does not differ substantially between systems with selective versus nonselective foragers, there is an important effect on long-term sustainability. Figure 5.4 depicts the average population density of predators in relation to prey carrying capacity and the slope of the relationship between energy concentration (e) and size (b). By and large, average population density of selective predators is substantially higher than that of nonselective foragers. In some cases, nonselective foragers are even driven extinct in systems capable of sustaining selective predators.

In previous chapters, we demonstrated important effects of predator behavior on population stability, because behavioral flexibility ensures that utilization is efficiently and sensibly directed toward parts of the resource spectrum that are momentarily advantageous, while ignoring other parts of the resource spectrum. Such behavioral refugia can exert a stabilizing effect on community dynamics in other contexts (see Chapters 2–4). Not so for size-selective predation. Instead, another fundamental dynamical feature is critically affected: population persistence. Foragers using size-selective strategies should be at a particular advantage in unproductive environments or systems with energetically poor prey.

The functional response of a nonfacultative forager is relatively simple, because β_i is equal to 1 for all i. For the adaptive forager, on the other hand, β_i varies depending on rank on the ith size class relative to the optimal diet breadth. The rate of consumption of size class i occurring at density N_i is dependent on densities, handling times, and probabilities of attack of all other size classes as well. For the same reason, the total number of prey consumed also depends on prey size composition as well as total density. It is therefore impossible to graph consumption rates for every possible situation that might arise. As an alternate approach, we therefore recorded total rates of prey consumption and intake of each size class for facultative and nonfacultative foragers during numerical simulations. We used a large number of size classes ($I = 20$) in these simulations in order to approximate a continuous size distribution. A sufficiently wide range of size classes also facilitates the self-thinning effects. The parameters of the model were such that prey density varies over a wide range. We then plotted intake rate against prey density, for the entire prey population (Fig. 5.5) or a single size class (Fig. 5.6).

The functional response is fuzzy at best in a system with this many dimensions, because the functional responses are two-dimensional projections of several interacting dimensions (the predator density plus all size class densities of the prey). Total consumption at any point in time is determined by the size composition of the entire prey population. The relative frequencies of these size classes vary from time to time in a dynamical system, muddying the resulting functional response. Note, however, that the functional response in such a size-structured system retains the basic positive decelerating shape of the simple disc equation (Fig. 5.5).

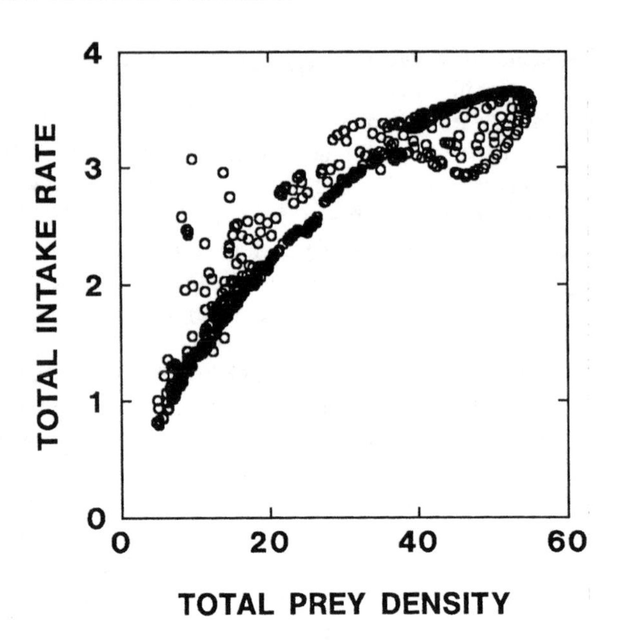

Figure 5.5. The predator functional response (per capita consumption rate) in relation to total prey density for a facultative forager. For a given prey density the rate of consumption is ambiguous because it is determined not only by resource density per se, but also on the actual composition of the entire resource population. As depicted here, it is a two-dimensional project of a k-dimensional manifold (k being the number of size classes, here $k = 20$). The general increasing but decelerating shape of the two-dimensional response is nevertheless retained. Parameter values ($K = 300$, $r = 1.0$, $b_{max} = 10$, $a = 0.025$, $\lambda = 0.2$, $d = 0.9$, $c = 0.2$, $\alpha = 0.1$, $m = 0.9$) were set such that the system shows sustained oscillations allowing for a wide range of resource density values.

This is no longer the case when considering intake of a single size class (Fig. 5.6) by a selective forager. Here, the predation rate is often close to zero over a wide density range, due to attention being directed unpredictably toward other size classes. At low prey densities, the predation rate on a single class can cover a wide range, once again dependent on the current composition of the prey population. Ambiguity in size-specific intake rates arises from the fundamental fact that diet breadth (in terms of size classes) clearly cannot be constant at any given prey density (Walton et al. 1992). We note in passing that if functional responses such as Figures 5.5 and 5.6 are unavoidable in our simple deterministic model, it is unlikely that patterns would be any cleaner in the real world having similar ecological relationships. Hence, it may prove difficult to interpret field data on consumption of size-structured resources without detailed knowledge of decision rules by the forager as well size class composition of the resource population.

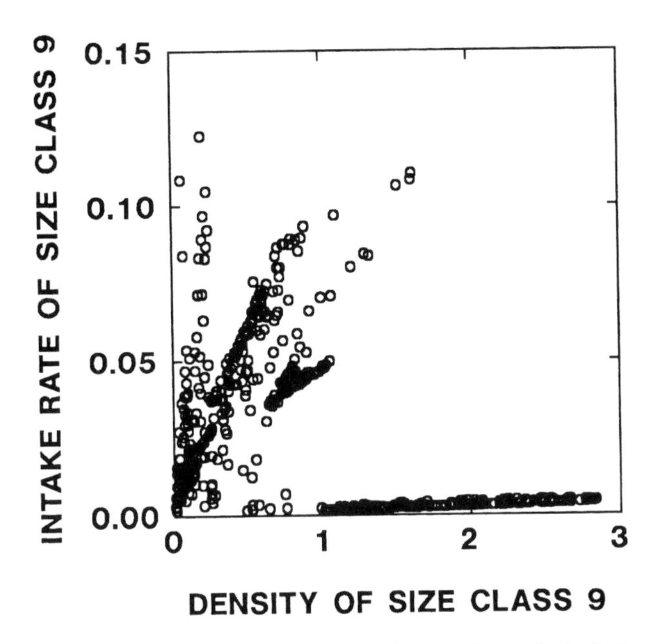

Figure 5.6. The rate of consumption of size class 9 (out of a total of 20 size classes) in relation to total prey density for a facultative forager. Parameter values as in Figure 5.5.

5.2 PARTIAL PREDATION MODEL

5.2.1 Consumption Model

In the foregoing, we have assumed that the predator can only consume a fixed proportion of each prey item, such as many plant–herbivore systems. We will continue this line of thought, adding another decision variable: partial predation.

It has been suggested (Lundberg and Åström 1990a,b; Åström et al. 1990) that herbivores feeding on relatively large and well-defined plant individuals (or, more generally, ramets) should view those plants as patches and exploit them accordingly. The marginal value theorem (Charnov 1976b) offers a general solution for such problems (see Chapter 4). It was early noted that this was true for any predator that sucks the juices of its prey. For example, Sih (1980, but see also Cook and Cockrell 1978; Giller 1980; Formanowicz 1984), showed that water-boatmen (*Notonecta*) had giving-up times of captured and partly ingested prey consistent with the marginal value theorem. The same is true for many nectar-feeding animals. Pyke (1980) tested the marginal value theorem on hummingbirds and Hartling and Plowright (1979), Hodges and Wolf (1981) and Hodges (1985) showed that bumblebees utilize flowers at least qualitatively in accordance with the predictions of the optimal patch use model.

Here, we make use of the prey-as-patch metaphor to evaluate how it works in a situation with a predator feeding in size-structured prey populations. Before deriving the optimal use of differently sized prey items, we must have some idea of how the removal of prey biomass works. We let the gain function specify how much biomass is removed (consumed) as a function of the time spent handling each prey. Handling time is now a variable under direct behavioral control, rather than an immutable parameter (Abrams 1990b). The predator can choose to spend a longer or shorter time handling prey of different size classes in order to acquire maximum energy gain. Assume that net energy gain is a monotonically decelerating function of handling time. Let us call this gain function G, calculated in the following way:

$$G_i = \frac{p_i b_i h_i e_i u}{p_i b_i + u h_i} \tag{5.10}$$

The parameter u determines how fast this function approaches the asymptote $e_i p_i b_i$, which is the maximum available energy available from size class i, and h_i, is the handling time of plants of the ith size class. In the previous models of size selection, we made handling time a linear function of prey mass, leaving no room for behavioral adjustment. Now, we will let handling time dictate both which size classes to include in the diet as well as how much mass to extract from each prey that is attacked.

Åström et al. (1990) discuss in detail the properties of different gain functions. We would emphasize two important biological assumptions. Our gain function assumes that the first few bites are as easily attained for a small prey as for a large one, even though both the extraction rate specified by u and the asymptote $p_i b_i$ may differ between size classes. There seems to be strong justification for that assumption, at least in plants (Åström et al. 1990). For this reason we have chosen a gain function like equation 5.10. Our gain function also assumes that even infinite handling time does not allow complete extraction of available energy from a single prey item, due to the asymptotic nature of the equation. This assumption is not readily defensible, but probably irrelevant unless prey densities become extremely small. It is nevertheless important to remember that the exact nature of the gain function (and its mathematical formulation) is critical. Åström et al. (1990) showed that small changes in the assumptions of the biological process producing the gain function can dramatically alter the predicted relationship between optimal handling time and patch (tree) size. Use gain functions with care!

To solve the plant consumption problem, we have to derive the optimal handling time for an average-size plant. We first need to estimate the expected time between encounters with prey items while the forager is searching. If we assume that prey items are randomly distributed in the habitat and that the predator also searches randomly, expected search time = $1/(aN)$, where a is area searched per unit time and N is total prey density. Our task is to find a unique optimal handling time for every size class in a multisize population. If we let the subscript a denote the average size of prey in the habitat, then the marginal value theorem (Charnov 1976*b*) tells us that

$$\frac{dG_i}{dh_i} = \frac{G_a^*}{\frac{1}{aN} + h_a^*} \tag{5.11}$$

should be true for all i under the optimal strategy, i.e., that the slope of the gain function for size class i (dG_i/dh_i) is equal to the average intake rate in the habitat (the right-hand side of eqn. 5.11). Before we can use equation 5.11 to solve for the optimal h_i we must know the quantities G_a^* and h_a^*. If we use equation 5.10 and let G and h be substituted by G_a and h_a, respectively (i.e., the gain function and the handling time for the average size class), then it is relatively easy to show that

$$h_a^* = \sqrt{\frac{b_a p_a}{auN}} \tag{5.12}$$

is the optimal handling time of average-size prey. By substituting h in equation 5.9 with h_a^*, we can also easily obtain G_a^*. Equation 5.11 is now possible to use for the calculation of the optimal handling time for every prey size i. We substitute the left-hand side of equation 5.11 with the derivative of equation 5.9 with respect to h, and use the obtained G_a^* and h_a^* expressions from above. The optimal handling time of size class i then becomes

$$h_i^* = \frac{b_i p_i}{u}\left(\sqrt{\frac{ue_i}{G_a^*}} - 1\right) \tag{5.13}$$

Equation 5.13 shows how the behavioral decision variable, the handling time of individual plants of size i, varies with attributes of both the size class itself and the rest of the foraging environment (through G_a^*). Handling times cannot be negative, so equation 5.13 will only be true if the square-rooted term in equation 5.13 exceeds 1. If it is ≤ 1 then h_i^* will be zero, i.e., a prey item of size i for which this is true should not be attacked. This is a result that traditional patch use models do not predict; they say that all patches should be visited regardless of size. Only the handling time differs between them (Stephens and Krebs 1986). Equation 5.13 therefore specifies both which size classes should be included in the diet and how much should be extracted from them, i.e., both diet selection and patch residence time are calculated simultaneously. The partial predation model predicts that larger individuals (large b_i) should be used for longer times than smaller ones, and that the richer the habitat (large G_a^*), the shorter the handling time per plant. Our main concern here is not, however, the optimal handling time per plant, but the consumption from plants of different sizes. The energy gain under the optimal policy is easily calculated by inserting equation 5.13 into equation 5.10 to produce G_i^*.

Before we proceed, we should say a few words about the term specifying the travel time between prey items. Although plants, for example, are often randomly distributed in a given habitat, encounters with them are not necessarily random. Åström et al. (1990) noted that moose (*Alces alces*) feeding on trees in winter tended to attack nearest-neighbor plants in the habitat. In a more detailed study, Gross et al. (1993) thoroughly measured movement patterns of foraging browsers in various spatial arrangements of plants. They also concluded that the spatial pattern of attack was not random, but reflected nearest-neighbor search. Lundberg and Danell (1990) modified the search time term based on an experiment in which plants were not randomly but regularly distributed in the habitat. These modifications may be correct in some circumstances, but the mathematics become even more complicated than the simple random model we have presented here. Deviations from random search could, however, be important for the dynamical properties of specific systems. Such effects should be considered when deriving appropriate foraging models.

Both the probability of attack and the biomass extracted once attack has occurred are intrinsic components of the partial predation model. The number of individuals of size class i that are attacked per time is

$$X_i = \frac{aN_i}{1 + a\Sigma h_i^* N_i} \tag{5.14}$$

Although superficially similar to the disc equation, the functional response specified by equation 5.14 is an ever-increasing function of prey density (Fig. 5.7), as for alternate forms in the literature (Lundberg and Åström 1990*b*; Lundberg and Danell 1990). This ever-increasing function tends to reduce the destabilizing inverse density dependence inherent in the disc equation (see Chapter 1). Another important feature of this functional response is that the number of prey taken per time unit is a negative function of prey size. The larger that average prey are, the more time the predator spends feeding on each food item and the fewer individuals attacked. In terms of prey mortality rate, this results in an interesting balance between the number of prey taken and the risk of dying once attacked (which is positively related to handling time).

The utility of treating the individual prey items as patches has long been recognized (Cook and Cockrell 1978; Sih 1980; Formanowicz 1984; Lucas 1985). The first explicit test on herbivores feeding on plants was done by Åström et al. (1990) and Lundberg and Danell (1990), using moose and tree stands in which all plants were the same size. As in most tests, however, Lundberg and Danell

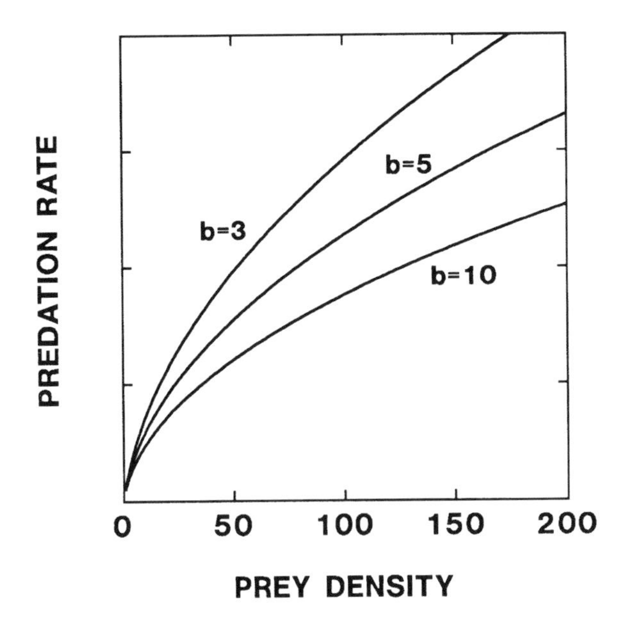

Figure 5.7. The functional response (the number of prey attacked per time unit) of a predator that forages according to the optimal partial predation model. The functional response is monotonically decelerating, but is not asymptotic. The functional response is calculated for single-sized prey populations, here shown for three different sizes (b is equal to 3, 5, or 10). If the prey individuals are large, the instantaneous rate of attack on prey is lower. The following parameter values were used: $a = 1$, $\lambda = 0.2$, e_i = constant = 1, $u = 1$.

(1990) used simplistic plant stands in which all plants were identical. Nevertheless, the experiments by Lundberg and Danell (1990) showed that the partial predation model was effective in predicting foraging time per tree as well as total intake rate per herbivore. Both the number of trees attacked as well as the biomass consumed in these experiments were well described by the optimal partial predation model; certainly better than three alternative functional response models. The main reason for this is that the optimal partial predation model takes into account the fact that handling time changes with both prey density and prey size composition (Giller 1980; Abrams 1982; Juliano and Williams 1985; Colton 1987; Lundberg 1988). Most other functional response models simply do not take these characteristics into account. The proportion of the available twig biomass per tree removed by moose was also well fitted by the partial predation model, the only one that simultaneously predicts both the number of prey attacked per unit time as well as the biomass extracted. Thus, there seems to be at least some empirical support for the "prey as patches" model.

5.2.2 Partial Predation and Population Dynamics

Before we can proceed to analysis of the full dynamical system with partial predation, we have to calculate the appropriate interaction terms for predators and prey. The forager's numerical response is calculated as $c\Sigma X_i G_i^*$, which is the number of prey items attacked multiplied by the energy gain from each prey item summed over all size classes. The parameter c is the coefficient converting consumption of prey mass into new foragers. Once a prey item has been attacked, we assume that the probability of death is directly proportional to the fraction of individual biomass consumed. Thus, the expression $mG_i^*/e_i b_i$ is prey mortality caused by a predator following the optimal partial predation strategy.

The nonfacultative strategy here is quite simple; take one-half the available biomass, such that $G_i = e_i p_i b_i/2$. As in the case of the optimal forager, prey mortality is directly proportional to the fraction of the prey biomass that is actually consumed. The probability of dying after an attack therefore equals $mp_i/2$. Substituting the appropriate terms in equations 5.7–5.9 with the new functional and numerical response terms, we are now in a position to analyze the dynamic properties of partial consumption. Changes in prey and predators using the optimal partial predation strategy are modeled by the following equations:

$$N_1(t+1) = \sum_{i=2}^{I} r\left(\frac{b_i}{b_{max}}\right) N_i(t)\left(1 - \sum_{i=1}^{I}\frac{b_i p_i}{K}\right) + P(t)\sum_{i=1}^{I} X_i\left(1 - \frac{mG_i^*}{e_i b_i}\right)S_{ij} - \frac{X_1 mG_1 P(t)}{e_1 b_1} \tag{5.15}$$

$$N_{i+1}(t+1) = N_i(t) - \frac{X_i mG_i^*}{e_i b_i}P(t) + \sum_{j=i+1}^{I} X_j\left(1 - \frac{mG_j^*}{e_j b_j}\right)P(t)S_{ij} \quad \textit{for } i \neq 1 \tag{5.16}$$

$$P(t+1) = P(t)\left(1 + \sum_{i=1}^{I} cX_i G_i^* - d\right) \tag{5.17}$$

where S_{ij} is the probability (either 0 or 1) that attacked individuals in size class j shrink to size class i. Changes in the null model with prey consumed by predators using the strategy "consume ½ the available biomass" are modeled by the following equations:

$$N_1(t+1) = \sum_{i=2}^{I} r\left(\frac{b_i}{b_{max}}\right)N_i(t)\left(1 - \sum_{i=1}^{I}\frac{b_i p_i}{K}\right) + P(t)\sum_{i=1}^{I} X_i\left(1 - \frac{mp_i}{2}\right)S_{ij} - P(t)X_1\left(\frac{mp_1}{2}\right) \quad (5.18)$$

$$N_{i+1}(t+1) = N_i(t) - P(t)X_i\left(\frac{mp_i}{2}\right) + P(t)\sum_{j=i+1}^{I} X_j\left(1 - \frac{mp_j}{2}\right)S_{ij} \quad \textit{for all } i \quad (5.19)$$

$$P(t+1) = P(t)\left(1 + \sum_{i=1}^{I}\frac{ce_i X_i b_i p_i}{2} - d\right) \quad (5.20)$$

When energy gain and prey mortality are changed according to a partial predation scenario, the predator–prey dynamics change considerably. The optimal strategy no longer ensures consumer persistence in the system. On the contrary, the nonfacultative strategy allows consumer persistence over a much wider range of parameter values. Figure 5.8 illustrates how mean consumer population density varies with prey carrying capacity and the slope of the relationship between prey size and energy content. The adaptive partial predation strategy allows persistence for only some of the selected parameter combinations, whereas the nonadaptive strategy persists for all. Population variability is lower for the adaptive forager than the nonadaptive one (Fig. 5.9). Hence, the effect of adaptive partial predation is similar to that of several other behaviors examined in this book: facultative foraging decisions tend to result in more stable community dynamics.

This outcome is not self-evident, given the fact that optimal patch use has a destabilizing component that has been rarely discussed in the literature (Lundberg and Åström 1990*b*, see also Sih [1984] for a discussion on foraging time and density-dependent predation). As resource density increases, the optimal handling time decreases. Therefore, the proportion of the available biomass that is eaten also decreases, reducing the mortality risk of plants. Prey mortality therefore decreases as plant density increases, everything else being equal. This means that herbivory contributes to inversely density-dependent mortality of plants, which is usually destabilizing.

The observed stabilizing effect of facultative partial predation is therefore surprising. We tentatively suggest that the answer may lie in asynchronous density variation among different size classes. One way to visualize the size-dependent dynamics is to plot the effect consumption has on the synchrony of density fluctuations of different size classes. We calculated cross-correlation coefficients

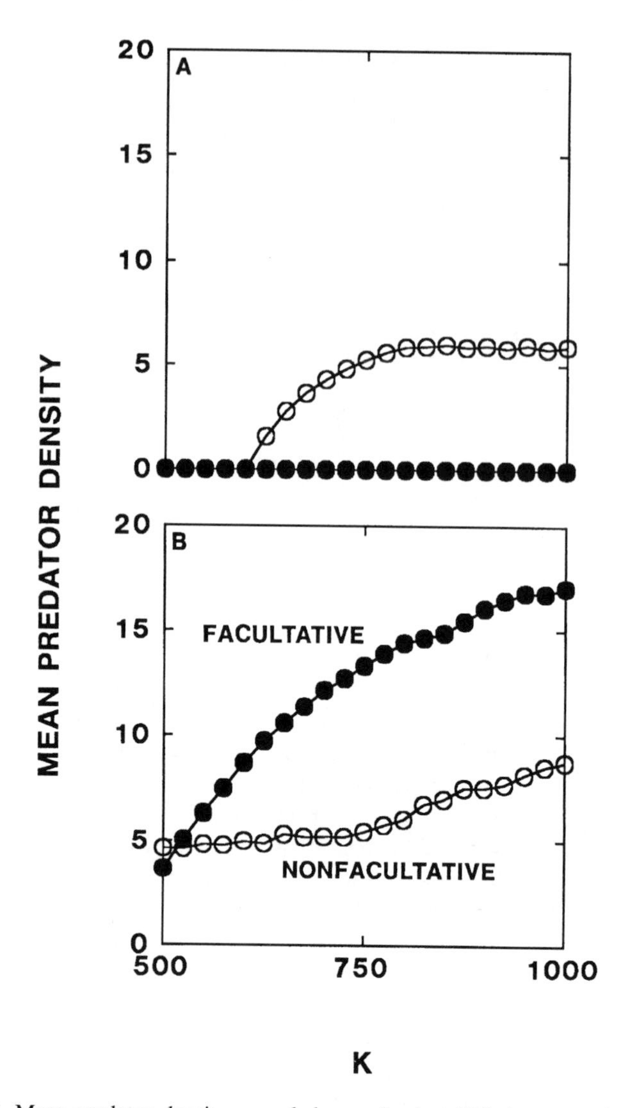

Figure 5.8. Mean predator density recorded over the last 500 time steps in 1000 time step simulations as a function of prey carrying capacity K for nonfacultative (open circles) and facultative (filled circles) predators according to the partial consumption model. Mean population densities are shown for different values of q, the slope of the relationship between energy concentration and size (A, $q = -0.1$; B, $q = 0.1$). Other parameter values as in Figure 5.3 with the additional parameter $u = 1.0$.

between pairs of size classes with lags between 1 and 20 time steps. From this matrix of cross-correlations, we then calculated the mean cross-correlation coefficient for each time lag, as illustrated in Figure 5.10. We note two things. First, the synchrony of plant density variation changes over time. Second, optimal

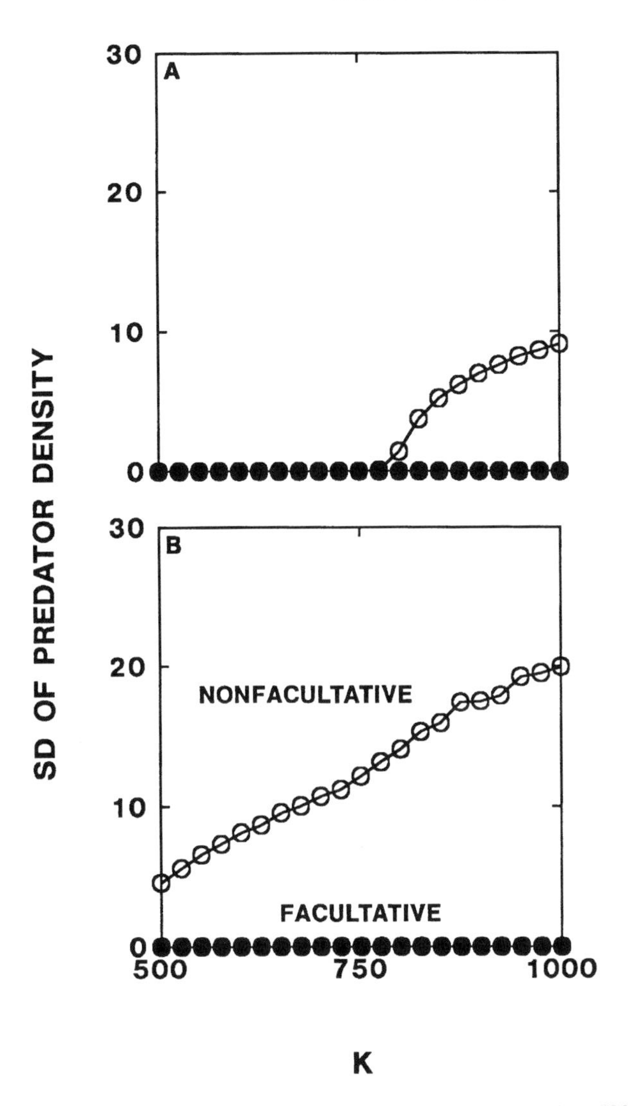

Figure 5.9. Standard deviation of predator density recorded over the last 500 time steps in 1000 time step simulations as a function of prey carrying capacity K for nonfacultative (open symbols) and facultative (filled symbols) predators according to the partial consumption model. Standard deviations are shown for different values of q, the slope of the relationship between energy concentration and size (A, $q = -0.1$; B, $q = 0.1$). Other parameter values as in Figure 5.8.

partial predation tends to exaggerate this asynchrony. Different size classes rarely fluctuate in phase. Asynchrony among size classes may tend to stabilize dynamics, much the way that asynchronous local dynamics stabilize spatially structured prey (Chapter 4).

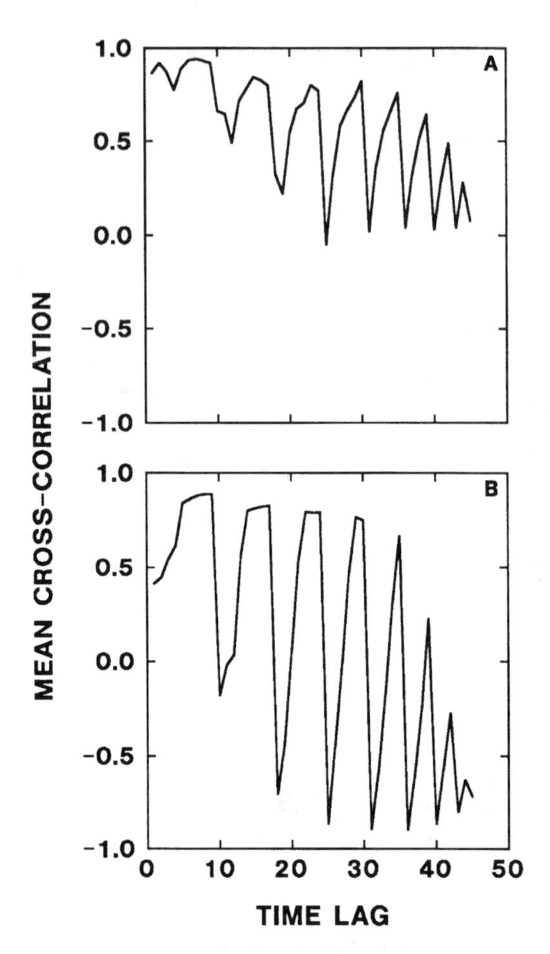

Figure 5.10. The change in mean cross-correlation coefficients over time for nonfacultative (*A*) and facultative (*B*) predators according to the partial consumption model. The mean cross-correlation is the mean of all pair-wise correlations between the 20 size classes of the prey population in time windows of 20 time steps. The *x* axis indicates the number of such consecutive time windows. High correlation means that all the size classes are fluctuating in synchrony, low correlation that they are asynchronous. Note that the degree of synchrony varies over time and that a system with a facultative predator tends to be less synchronous than a system with a nonfacultative predator. Parameter values as in Figures 5.3 and 5.8.

5.3 THE SIZE STRUCTURE CHALLENGE

Life-history theory and demography link ecological and evolutionary perspectives on population processes. Because, one way or another, size structure inevitably must be a part of a viable theory of populations, it has attracted a great deal of attention. The substantial literature on size-structured life-history theory bears witness to this. We refer the reader to major recent reviews for full treatment of the general issues (Metz and Diekmann 1986; Ebenman and Persson 1988; Caswell 1989; Roff 1992; Stearns 1992). Although size structure has a long pedigree in life-history theory, few attempts have been made to combine detailed size-related behavioral mechanisms and long-term population dynamics. Physiologically structured models have recently come into vogue as a means of handling properties of both individual members of the population and the population itself (Metz and Diekmann 1986; Metz et al. 1988; Caswell and John 1992; DeAngelis and Rose 1992; Persson et al. 1996). Both crustaceans (Metz et al. 1988; De Roos et al. 1992) and fish (e.g., DeAngelis et al. 1991; Rice et al. 1993; McCauley et al. 1993; Tyler and Rose 1994) populations have been analyzed with this kind of model approach. Individual-based models are excellent for modeling variability among individuals and behavioral decisions over short time scales but are sometimes awkward for modelling long-term dynamics. Physiologically structured models based on diffusion processes (e.g., Metz and Diekmann 1986) can do the job of projecting long-term dynamics, but are less adequate for combining stage-structured demography (like our size distributions with self-thinning) with behavioral decision making. The models we have dealt with in this chapter are intermediate on the continuum between the detailed individual-based models and diffusion models.

No matter what stand we take in relation to these approaches, there is an inevitable price to pay: lack of insight into mechanisms by which demography, behavioral decisions, population interactions and evolution eventually enhance or diminish persistence and stability. It is obvious, however, that we have to take size structure seriously. Size-selective predation is both widespread and fundamentally important for virtually all animal taxa studied, including insects (Charnov 1976*a*; Pastorok 1981), gastropods (Hughes and Burrows 1991), crustaceans (Elner and Hughes 1978), fish (Werner and Hall 1974; Stein 1977; Bence and Murdoch 1986; Paszkowski et al. 1989; Persson 1991), birds (Goss-Custard 1977; Cayford and Goss-Custard 1990), and mammals (Fryxell and Doucet 1993). Although the list of size selection experiments is long, there are not many examples of studies in which more crucial aspects have been measured or calculated. We just report a few that show results that have a direct bearing on the mechanisms of size-structured predator–prey dynamics.

In a study on cannibalistic feeding behavior of the hemipteran *Notonecta*, Streams (1994) showed how functional responses and per capita rates of prey mortality were actually modified by prey size selectivity. The functional responses derived for two size classes of prey (smaller size classes of *Notonecta* itself!) did not seem to differ substantially. The demographic effect of size-structured predation was pronounced. Per capita mortality rates were density independent for small prey but became inversely density dependent for large prey.

Osenberg and Mittlebach (1989) studying sunfish (*Lepomis gibbosus*) predation on gastropods experimentally derived how the probability of attack should vary with prey energetic value, which in this case corresponded to prey size. We show one of their results in Figure 5.11, which nicely underpins one of our basic assumptions behind the size selection model. Not only should we expect prey sizes to vary in profitability, but also attack rates should covary with energetic profitability. The observed continuous variation in attack probability with prey size is consistent with our β function.

The functional response experiment with beavers (*Castor canadensis*) by Fryxell and Doucet (1993) also gives insight to the importance of size structure and partial predation for the plant–herbivore interaction. In a series of trials, beavers were offered saplings of 3 species of trees. In single-species, single-size trials, the functional response by beavers was a monotonically increasing and decelerating type II curve. When several size classes were offered to the animals, they exhibited diet choice in the sense that larger saplings had a higher probability of attack, at least when overall sapling density was high. Since beavers spend a longer time handling larger trees, the maximum rate of cutting is depressed. At low sapling densities, smaller saplings were included in the diet and mean handling time was reduced and cutting rate increased. In all cases they studied, the probability of attack was related to sapling profitability. As predicted by classic optimal foraging theory, this directed attack on most profitable prey was most pronounced at high sapling densities.

No matter how much evidence for size-selective predation we accumulate, there is still a nagging need to better explain and understand what is happening

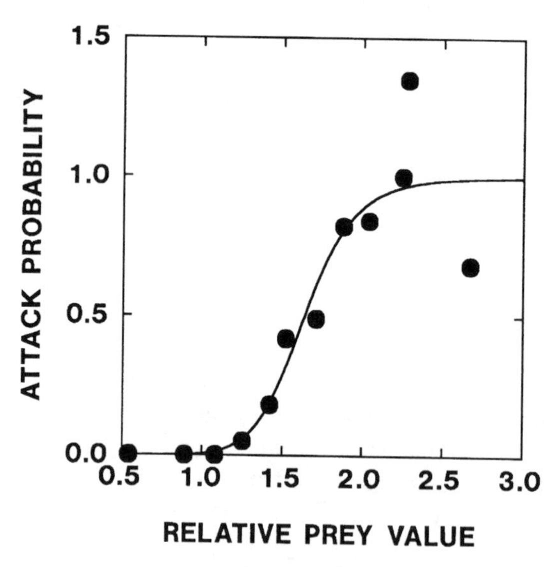

Figure 5.11. The probability of attack in relation to relative prey value (corresponding to size class) for pumpkinseed sunfish (*Lepomis gibbosus*) feeding on lake gastropods. These results show that there is a strong energetic reason for prey size-selectivity and the fish evaluated the size range of prey corresponding to the β-function we have been using throughout. Data from Osenberg and Mittlebach (1989).

in these high-dimensional systems. We think there may be two major reasons for the rather enigmatic results from our analyses.

First, density-dependent effects of size-selective foraging behaviors may be masked by cycles induced by size structure in the resource population. For example, Ruxton et al. (1992) showed that age-structured prey populations had very little response to interference among predators, unlike those in nonstructured models. This is because there is a balance between "prey-escape cycles" caused by relatively slowly reacting predators to fast growing prey (allowing prey to escape heavy predation and to grow to high densities) and "generation cycles" (Gurney and Nisbet 1985) caused by within- and between-generation competition and mortality. For many parameter combinations the "generation cycles" are immune to density-dependent mortality effects (reducing the "prey escape cycles") caused by predation. It is difficult to disentangle such effects in our systems with certainty. Our self-thinning rule probably contributes to generation cycles. Ruxton's et al. (1992) model had four components (food, prey larvae, prey adults, and predators) in contrast to our 20 prey size classes. This would lead to increased time delays and even more pronounced generation cycles for our model.

Second, the effects of size structure on the dynamics may not come into play unless there is also spatial structure. It is well established, particularly in fish, that size structure is often accompanied by size-related habitat shifts ("ontogenetic niche shifts," Werner and Gilliam 1984; Persson 1987, 1988; Werner and Hall 1988). Intraspecific competition for food, possibly also cannibalism, and predation risk may be balanced in a manner leading to habitat segregation. The size-structured population thus becomes simultaneously spatially segregated with a one-way transition of smaller individuals from one habitat to another as they mature. If so, then there is a strong potential for predator-free refuge effects, with dynamical implications we consider in Chapter 4. Although ontogenetic spatial segregation is well established (e.g., Persson et al. 1996 for review), the implications for long-term population dynamics have only begun to be explored (Werner 1992; Werner and Anholt 1993; Persson et al. 1996).

Increased dimensionality of size-structured populations inevitably makes both empirical and theoretical work considerably more difficult, regardless of the models one chooses to use. Despite the somewhat ambiguous results, we have nevertheless equipped ourselves with some useful tools for further explorations of consumer-resource systems with size structure. Self-thinning may be universal, whereas partial consumption is not, but we feel that the plant–herbivore systems for which those characteristics are true deserve more attention. One might even argue that we know least about plant–herbivore relationships, which are the very basis for most other food-web interactions.

5.4 SUMMARY

Size variation is a fundamental fact of life, yet the ecological implications of size variation are far from clear. In this chapter, we ask how differential fecundity and risk of mortality among life stages may influence trophic interactions. These life history characteristics are often strongly associated with body size. We therefore develop a predator–prey model in which the prey population is size structured. We make the well-substantiated assumption that intraspecific competition for limited resources among the prey leads to self-thinning. The size distribution interacts with density in such a way that increased average size forces the prey density to decline. Each size class is associated with its own probability of reproduction, competition with conspecifics, and risk of being attacked by predators.

Using this model, we examine two scenarios. First, we let the rate of attack by predators be contingent on prey size and prey energetic profitability. Prey mortality is a function of the proportion of prey body mass available for consumption. Second, we extend this to let the forager regard each resource individual as a patch, letting food extraction from individual prey become a function of the profitability of other "patches."

The dynamic effects of adaptive versus nonadaptive forms of foraging does not differ as clearly for size-structured populations as it did in previous chapters. Partial prey consumption does tend to have a slight stabilizing influence on dynamics, whereas nonlethal consumption without partial predation only affects forager persistence. We discuss these findings in relation to recent theoretical and empirical work on the interaction between generation cycles within the resource population and predator-induced prey-escape cycles, as well as the problem of spatially segregated stage-classes in the prey.

6 Interference and Territoriality

Theoretical models of trophic interactions often assume that predator and prey populations have overlapping spatial distributions and that predator dynamics are dictated solely by prey abundance. Although these circumstances may sometimes be true, such cases are probably the exception rather than the norm, particularly for social species of predators that either fight directly with each other or else territorially exclude other predators from gaining access to food resources. In this chapter, we consider the ecological ramifications of direct interference among predators and the effect of territoriality on the spatial distribution of resources and resource use by fitness-maximizing individuals.

6.1 INTERFERENCE

Models of behavioral interference between predators have been derived several times (Hassell and Varley 1969; Beddington 1975; DeAngelis et al. 1975; Sutherland and Parker 1985, 1992; Korona 1989; Ruxton et al. 1992; Moody and Houston 1995; Holmgren 1995). These behavioral models share a common assumption that the frequency of agonistic interactions among predators depends on the rate of predator movement in the environment, for the same reason that movement rates directly influence the predator functional response. A useful way to visualize such aggressive interactions is to set up a dynamical system of 4 classes of predators (Ruxton et al. 1992): those that are actively searching for prey (S), those that are handling or consuming prey (H), those that are currently engaged in fighting but were formerly searching for prey (W), and those predators that are currently fighting but were formerly handling prey (Z). The symbols depict the abundance of each subclass of predator at a specific point in time.

Assume that interactions can only be initiated by searching predators, but the recipients of such aggression would be otherwise engaged in other activities: searching, handling, or fighting. We know from our derivation of the functional response (Chapter 1) that searching predators will encounter prey at a rate proportionate to the area searched per unit time (a) multiplied by searcher density (S) multiplied by prey density (N). Hence, the abundance of searching predators will tend to decline over time by $-aSN$ as predators locate food items. The frequency of encounters between various classes of predator are similarly derived. Handling (H) or fighting (W or Z) predators are themselves stationary, so the abundance of searchers will change by $-aSH$, $-aSW$, or $-aSZ$, reflecting random encounters between a mobile searcher and immobile subclasses of predator. The frequency of encounters among searchers is squared because both searchers are mobile, hence searcher abundance would decline by $-2aS^2$. On the positive side of the ledger sheet, former handlers or fighters are converted to searchers according to H/h or W/w, where h is handling time and w is fight duration. One can similarly derive differential equations for all other predator subclasses:

$$\frac{dS}{dt} = -aSN - 2aS^2 - aSH - aSW - aSZ + \frac{H}{h} + \frac{W}{w} \tag{6.1}$$

$$\frac{dH}{dt} = aSN - aSH - \frac{H}{h} + \frac{Z}{w} \tag{6.2}$$

$$\frac{dW}{dt} = 2aS^2 + aSH + aSW + aSZ - \frac{W}{w} \tag{6.3}$$

$$\frac{dZ}{dt} = aSH - \frac{Z}{w} \tag{6.4}$$

Simulations of such systems, while holding prey density constant, invariably approach a stable equilibrium due to the linear form of the equations. The resultant proportion of searchers at behavioral equilibrium can be obtained by simultaneously solving these 4 equations (Ruxton et al. 1992):

$$\frac{S_{eq}}{P} = \frac{-1 + \sqrt{1 + \dfrac{4awP}{1 + ahN}}}{2awP} \tag{6.5}$$

It follows that the rate of prey consumption per predator equals aNS_{eq}/P. If predator density is small relative to that of prey ($P << [1 + ahN]/4aw$), then one can use the following close approximation for the functional response (Beddington 1975; Ruxton et al. 1992):

$$X = \frac{aN}{1 + ahN + awP} \tag{6.6}$$

This functional response equation predicts that consumption declines curvilinearly with predator density ($\partial X/\partial P < 0$), introducing a direct density-dependent term in the predator population response (Beddington 1975; DeAngelis et al.

1975). Following the groundbreaking work of Hassell and Varley (1969) on parasitoid interference, it has become conventional to measure the intensity of predator interference by the magnitude of the slope of log-transformed values of both intake and predator density (Sutherland and Parker 1985, 1992). The behavioral model derived above implies that interference (the slope of the log–log intake function) becomes stronger with increasing predator density (Fig. 6.1), asymptotically approaching a value of 1 as $P \rightarrow \infty$ (Beddington 1975; Moody and Houston 1995). Interestingly, a wide range of studies of predators as different as insect parasitoids (Hassell 1971), oystercatchers (Goss-Custard and Durrell 1987*a,b,* 1988), and migratory caribou (Manseau 1996) have recorded interference curves with accelerating slopes similar to that predicted. Similar density-dependent functional responses can be derived from alternate assumptions that only searching predators interact (Ruxton et al. 1992; Moody and Houston 1995) or that aggression results from kleptoparasitism directed toward successful predators (Holmgren 1995). The similarity of outcomes suggests that the behavioral details have relatively little importance; the crucial density dependence derives from the opportunity cost of competition with feeding time.

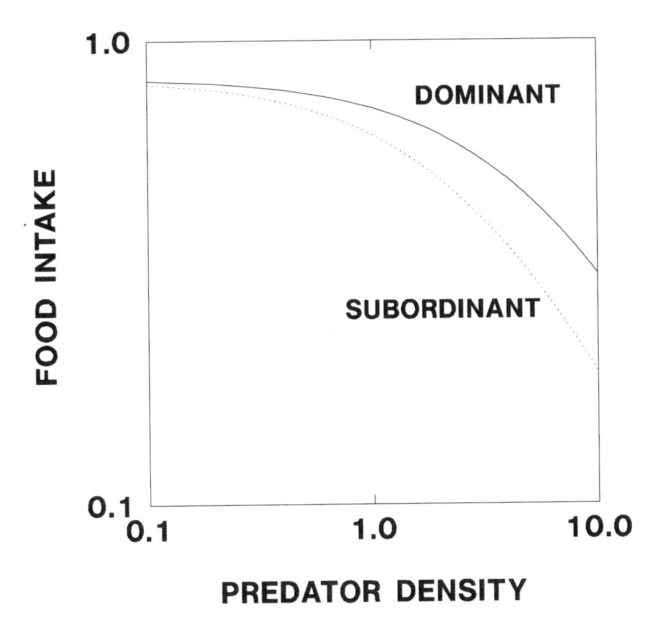

Figure 6.1. Rate of food intake by interfering predators as a function of predator density, while holding prey density constant ($N = 5$, $a = h = 1$, $w = 2$) plotted on logarithmic axes. The top curve in each diagram pertains to a predator with small opportunity costs due to interference because the individual is of high social rank, whereas the lower curve corresponds to a low-ranked individual.

6.1.1 Interference and Population Dynamics

It is straightforward to include predator interference in our now-standard predator–prey system:

$$\frac{dN}{dt} = rN(1 - N/K) - \frac{aNP}{1 + ahN + awP} \tag{6.7}$$

$$\frac{dP}{dt} = \frac{aceNP}{1 + ahN + awP} - dP \tag{6.8}$$

The outcomes of such dynamical simulations can be readily understood by plotting the time trajectories for both participants in the predator–prey phase plane. Negative density-dependence in predator growth rates due to interference with foraging imparts a positive slope on the predator zero isocline (Fig. 6.2), unlike the vertical isocline of pure Lotka–Volterra predation models. By the same token, predator interference skews the prey zero isocline, such that the hump of the prey isocline occurs at lower prey densities than would otherwise be the case. Both of these features improve the possibility that the intersection of the two isoclines occurs to the right of the hump in the prey isocline, at which point density-dependent (stabilizing) processes supersede the usual destabilizing effects of predation on prey net recruitment. As a consequence, systems with predator interference are often stable over a much broader range of parameter values (Fig. 6.3*B*) than their counterparts without interference (Fig. 6.3*A*). All the available flavors of interference models show similar stabilizing effects on trophic interactions (Hassell and Varley 1969; Hassell and May 1973; Beddington 1975; DeAngelis et al. 1975; Ruxton et al 1992). It is not yet clear, however, whether the cost of interference found at population densities typically found in nature would be sufficient to be stabilizing.

6.1.2 Social Structure and Interference Levels

We have assumed thus far that all predators are equally vulnerable to interference. This runs counter to a large body of empirical studies showing wide variation in social dominance status. One of the best-known examples of the effect of social dominance on feeding rates is the long-term work by Goss-Custard and colleagues on oystercatchers feeding on mussel beds in coastal England (Goss-Custard and Durrell 1987*a,b,* 1988; Goss-Custard et al. 1995*a,b*). Their work clearly shows that behavioral interference between birds affects rates of forage intake and the subsequent forager spatial distribution. Interference costs in oystercatchers vary across individuals, across spatial units, and across years. A full understanding of interference therefore demands consideration of the effects of variation in social status.

Following Goss-Custard et al.'s (1995*a*) logic we can expand Beddington's

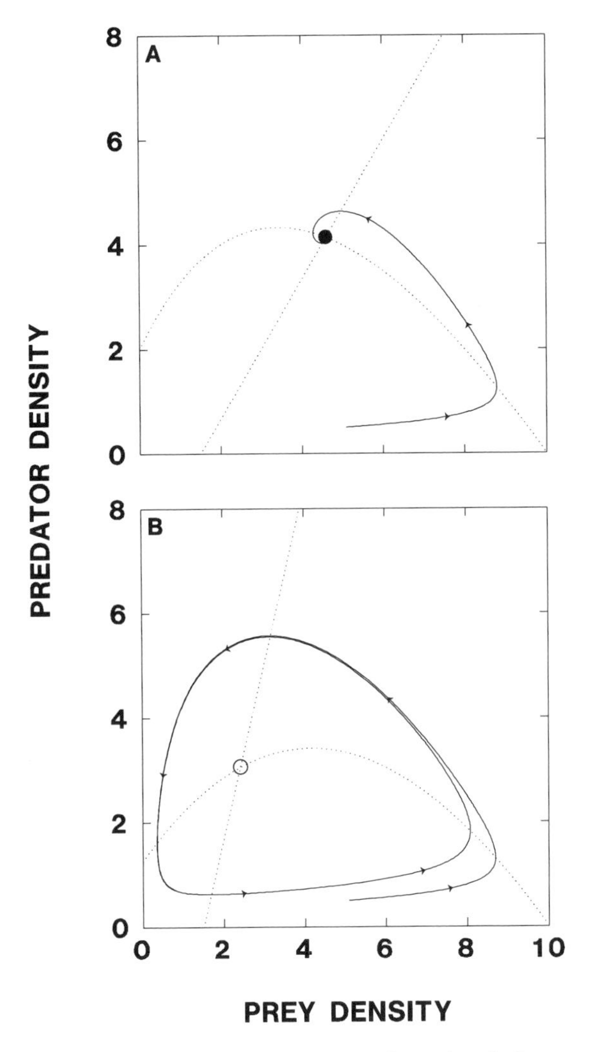

Figure 6.2. Trajectories over time of interfering predators in relation to prey, for large time costs for each encounter between predators (*A*) and small costs (*B*). The dotted lines indicate zero isoclines for prey (parabola) and predators (line) Nontrivial equilibria are indicated by symbols (filled = stable, open = unstable). ($a = h = r_{max} = c = e = 1$, $d = 0.6$, $K = 10$, $W = 0.5$ in *A* and $W = 0.2$ in *B*).

(1975) interference model to include social structure by weighting the frequency of fights by the probability that a given forager has lower social status than potential aggressors. Assume that dominance ranges between 0 (for the weakest individuals in the population) and I (for the most dominant individuals) and let β_i denote the cumulative probability that an encountered individual ranks lower

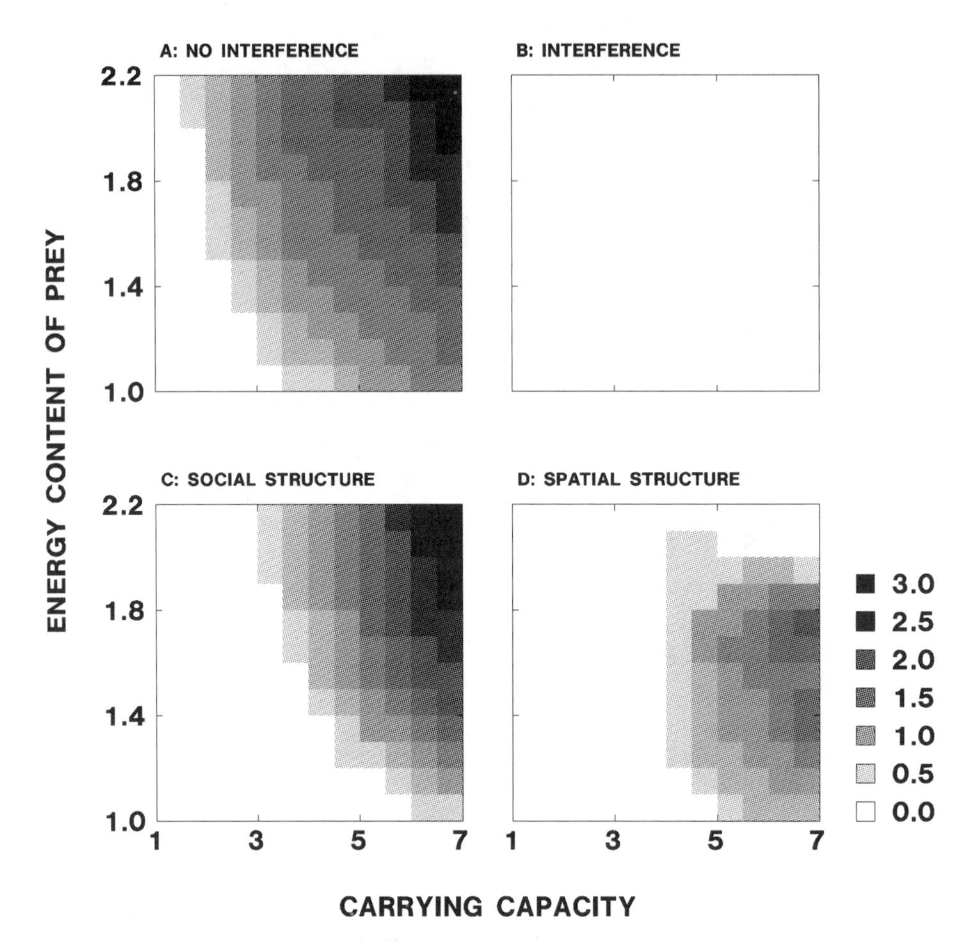

Figure 6.3. Variation in predator densities over last 2000 time steps of 4000 time step simulations for different parameter combinations for systems with noninterfering predators (*A*), simple interference (*B*), social dominance affecting probabilities of interference (*C*), and social dominance coupled with "ideal-free" habitat selection (*D*). Common parameters were used in each simulation ($w = 0.5$, $h = c = r_{max} = 1$ and $d = 0.6$ in all cases; $a = 1$ in cases *A*, *B*, and C; $a_1 = 1.5$ and $a_2 = 1.0$ in *D*, and normal distribution in f_i with mean $(i) = 10$ and $SD(i) = 2$ in cases *C* and *D*).

than a searcher of rank i. Wasteful interactions result from encounters with predators of higher dominance status (Goss-Custard et al. 1995*a*), the frequency of which scales according to $1 - \beta_i$. This results in the following interference model for the functional response of an individual of dominance rank i:

$$X_i = \frac{aN}{1 + ahN + aw(1 - \beta_i)P} \tag{6.9}$$

Outcomes of this simple model of social structure produce variation in intake at a given level of prey density, ranging from high intake for high-ranking individuals to low intake for low-ranking individuals (Fig. 6.1). Low-ranking individuals experience a more dramatic drop in food intake with increasing predator density than high-ranking individuals, but all individuals experience density-dependent effects on intake. Sutherland and Parker (1985) identified two alternate forms of social variation. Interference curves that have identical intercepts but different slopes were ascribed to "individual variability in foraging efficiency" whereas curves with identical slopes but different intercepts were ascribed to "individual differences in interference experienced." The functional response model we have described typifies variation in efficiency at low predator densities but variation in interference experienced at high predator densities (Fig. 6.1), suggesting that both patterns might occur in a given species, albeit at different densities.

We tested the effects of social structure on trophic dynamics by embedding the social functional response (eqn. 6.9) in the standard Lotka–Volterra formulation (eqns. 6.7 and 6.8 with appropriate modifications to functional and numerical responses, summing over the full range of social ranks). It is not entirely clear a priori which probability function would be most appropriate for β_i, so we tried three alternatives: uniform, normal, and gamma distributions. Regardless of the underlying frequency distribution, the models with social structure were always less stable than their counterparts without social structure (e.g., Fig. 6.3*C* for normally distributed dominance status versus Fig. 6.3*B* for a nonstructured model). In retrospect, this is not surprising because $1 - \beta_i$ is always less than 1, thereby diluting the probability of interference and its stabilizing properties.

6.1.3 Spatial Structure and Interference

Our social interference model is still somewhat simplistic, however, in assuming that individuals cannot reduce the risk of interference by moving into alternate habitats. Moody and Houston (1995) and Holmgren (1995) have elegantly shown that adaptive habitat selection by predators faced with interference rapidly leads to an "ideal-free" spatial distribution. Following their approach, we revised our basic interference model to include optimal habitat selection based on fitness-maximizing choice between 2 habitats with search efficiencies (the parameter a varying between habitats) à la Chapter 4. Top-ranking individuals chose habitats first, followed by second-ranked individuals, on down to the lowest-ranking individuals. Regardless of the parameter values used, our optimal habitat selection models produced semitruncated spatial distributions (Holmgren 1995), in which high-ranking individuals occupy the best habitat, with a variable mixture of lower-ranking individuals occupying both habitats (Fig. 6.4). In unstable systems,

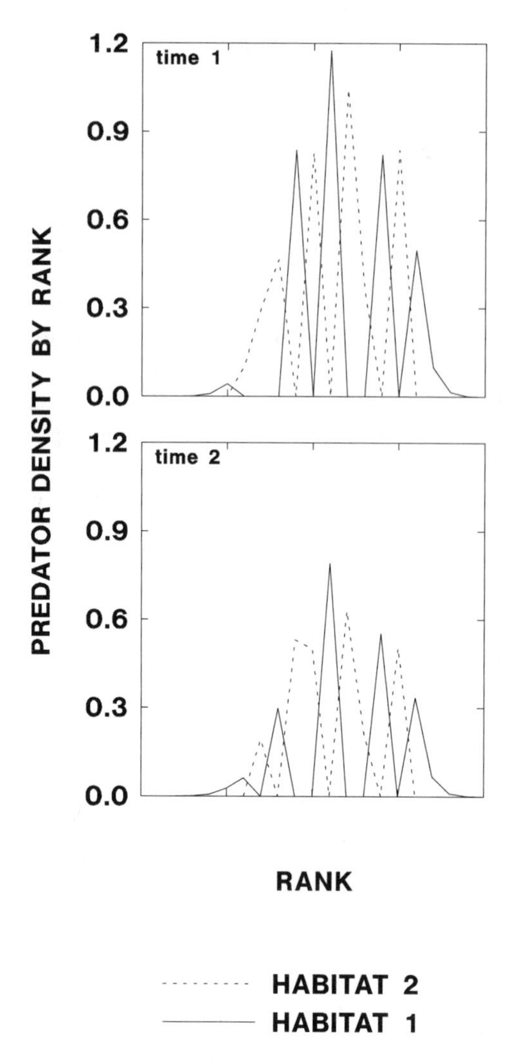

Figure 6.4. Habitat distribution at 2 distinct points in time for a temporally unstable population of interfering predators that have both social and spatial structure (a_1 = 1.5, $a_2 = h = r_{max} = c = e = 1$, d = 0.6, K = 10, W = 0.5). The solid line indicates relative proportion of individuals of given social rank present in habitat 1, whereas the dotted line pertains to individuals in the poorer habitat 2.

low-ranking individuals flip between habitats whereas dominants are always found in the better habitat (Fig. 6.4). Spatial heterogeneity improved stability of the interference model with social structure (Fig. 6.3*D*), although it was still not as stable as the simplest interference model without social or spatial structure.

In sum, our results lend support to earlier suggestions in the literature that interference can indeed stabilize trophic interactions. We add the caveat that this stabilizing effect is diminished by individual variability in social dominance, but it is enhanced by spatial heterogeneity. We now address another common form of social interaction, territoriality, which involves aspects of both behavioral interference and spatial refugia for prey.

6.2 TERRITORIALITY

6.2.1 Optimal Territory Size

There have been numerous treatments of the adaptive basis of territoriality, often derived from the same general principles (Dill 1978; Kodric-Brown and Brown 1978; Ebersole 1980; Hixon 1980; Schoener 1983, 1987; Davies and Houston 1984; Houston et al. 1985; Hart 1987; McNair 1987; Stephens and Dunbar 1993). If there are substantial trade-offs between the costs and benefits of defending a territory of given size, then in principle it should be possible to determine the optimal size of territory to defend in relation to changes in resource availability or predator density. We will simply extend this general approach to our simple Lotka–Volterra based system to evaluate the dynamical implications of territorial interference.

Virtually all treatments of territoriality assume that there are diminishing returns with increased territory size, i.e., fitness gains increase less than proportionately with territory size, possibly due to decelerating functional responses of territorial foragers due to processing constraints or else limits on the time available for territorial defence versus feeding. In either case, territoriality makes sense only if the carrying capacity of prey is limited, regardless of whether prey are plants or animals.

First consider the case in which time costs are trivial but predators have a decelerating functional response. One can predict the long-term benefit by first calculating the equilibrial prey density for a territory of given size (A). Imagine for the moment a situation in which 5 predators split up an island of 10 km^2 into equal territories of 2 km^2. If an additional 5 predators arrived on the scene, yet couldn't acquire territory space, the mean density within each territory would remain ½ animal per km^2. This example shows that the functional density of solitary predators within an exclusive territory = $1/A$. If we assume that prey show logistic population growth in the absence of predators and that predators have a type II functional response, then the rate of change of prey within a territory of size A can be calculated as

$$\frac{dN}{dt} = rN\left(1 - \frac{N}{K}\right) - \frac{aN}{A(1 + ahN)} \tag{6.10}$$

At equilibrium, $dN/dt = 0$, which in this case occurs when $1/A = a(r - rN/K)(1 + ahN)$. A bit of algebraic rearrangement yields the equilibrium prey density (N^*) in relation to territory size

$$N^* = \frac{-Ak_2 \pm \sqrt{A^2k_2^2 - 4Ak_1\,(Ak_3 - 1)}}{2Ak_1} \tag{6.11}$$

where $k_1 = rh/K$, $k_2 = (r/K)(hK - 1/a)$, $k_3 = r/a$. Provided that the territory exceeds a critical minimum size, $A_{min} = 4hK/r(hK + 1/a)^2$, then there are always two possible equilibrium prey densities (Fig. 6.5). The upper limb of the parabola is stable, implying that a slight perturbation away from N^* with no change in territory size would lead to recovery of N^*. In contrast, if by chance prey density ever falls below the lower limb of the parabola, then prey would be unable to sustain consumption and local extinction would inevitably follow, as seen for example in plant–herbivore models with constant herbivore stocking levels (Noy-Meir 1975). For simplicity, we assume that there are always sufficient prey to guarantee a stable prey equilibrium, such as would occur if a group of territorial predators invaded a pristine population of prey.

Because the rate of consumption increases with prey density, the equilibrial energetic benefit (B) (energy content multiplied by the functional response) increases with territory size, but at a decelerating rate. By the same token, one might expect the energetic costs (C) of defending a territory to increase with A. The form of this territorial cost function is debatable (Kodric-Brown and Brown 1978; Stephens and Dunbar 1993). If costs relate strictly to the length of the perimeter or to the average distance between spatial positions occupied by the

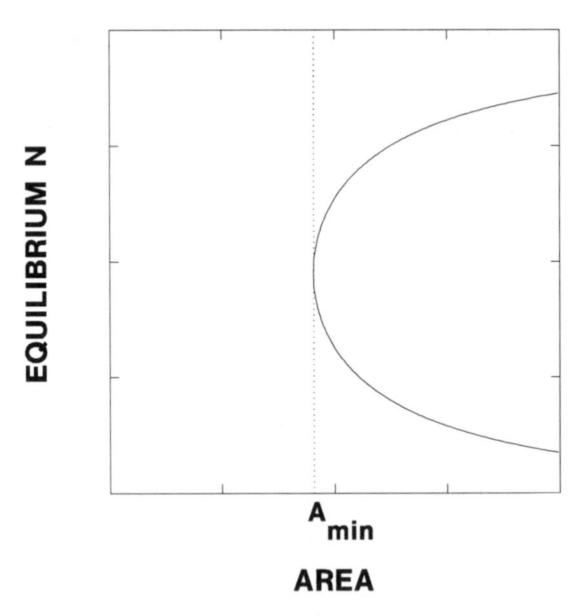

Figure 6.5. Equilibrium prey density (N^*) in relation to territory size of predators. A_{min} is the minimum territory size permitting a sustainable population of prey.

territory holder and an intruder, then $C \propto A^{1/2}$ (Stephens and Dunbar 1993). In contrast, one might predict from random search theory that the likelihood of an intruder encountering a given territory should be linearly proportional to A (Hixon 1980; Schoener 1983, 1987). The precise form of the C function matters little for our purpose, so we assume the simplest case: $C \propto A$. Regardless of the relationship between C and A, the cost of defense should be proportional to the number of intruders, because of changes in energetic expenditure or elevated risk of mortality during agonistic displays. Fitness (w) is a complex function of A, calculated according to

$$w = \frac{aceN^*}{1 + ahN^*} - (d + \theta AP) \tag{6.12}$$

where θ = the cost of defense per unit area. This fitness function has a single maximum value at intermediate territory sizes (Fig. 6.6), and the optimal territory size depends on the richness of the environment (K, because it influences N^*) and the density of predators potentially intruding (P)(Fig. 6.7). Note that the functional response is dictated by prey density, not absolute prey abundance summed over the entire territory, a subtle, but nonetheless important, logical inconsistency that has slipped advertently into several previous territoriality models.

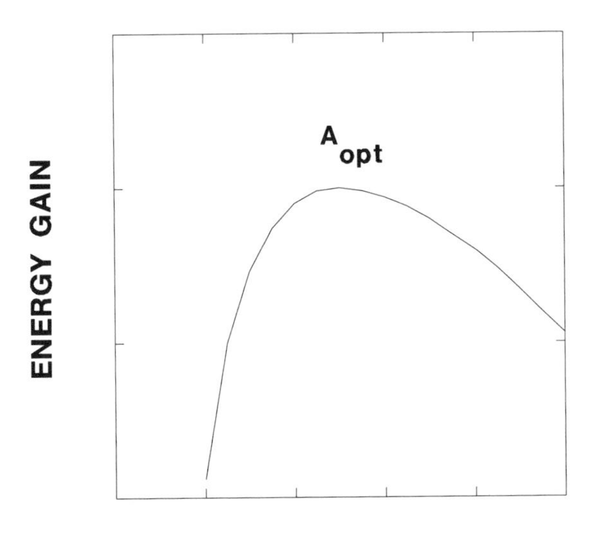

Figure 6.6. Equilibrial fitness of predators maintaining a territory of various sizes. The asymmetric fitness function generally has an optimal value at a territory of intermediate size.

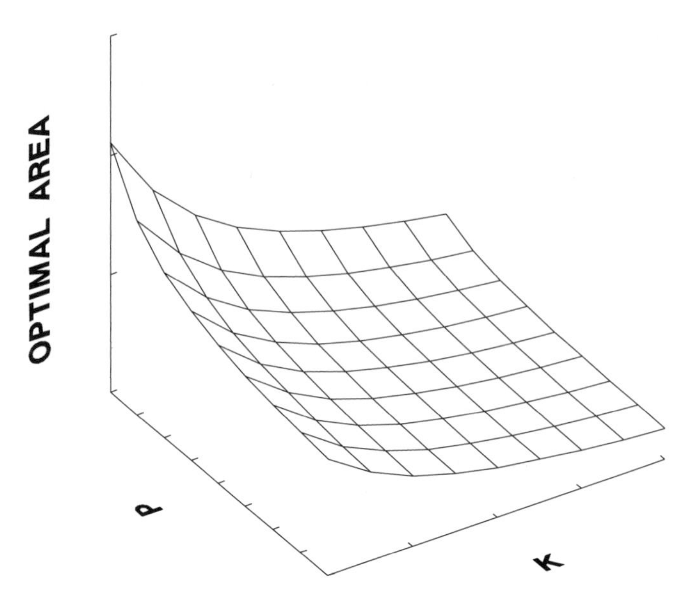

Figure 6.7. Optimal territory size in relation to predator density and carrying capacity of the environment.

6.2.2 Evidence for Optimal Territory Size

There is reasonably strong evidence that territory size is facultatively related to resource richness in some species. For example, Kodric-Brown and Brown (1978) and Gass et al. (1976) found that territory size in rufous hummingbirds was inversely correlated with flower density. Interestingly, the hyperbolic relation between territory size and flower density implies a constant level of resources per individual despite seasonal or spatial variability in floral abundance. Similar inverse correlations between resource richness (i.e., K) and territory size have been observed in a variety of species, including hummingbirds (Gass et al. 1976; Kodric-Brown and Brown 1978), shorebirds (Myers et al. 1979*a,b*), stream insects (Hart 1985), small forest mammals (Mares and Lacher 1987), and deer (Bobek 1977). Perhaps more importantly, in some of these cases, experimental manipulations of food availability induced predictable changes in territory size. Nonetheless, there are some examples of relationships between territory size and resource richness running counter to the model predictions. In one of the best-documented examples, Ebersole (1980) found that female damselfish increased territory size in response to experimental increases in resource richness, whereas males usually (but not always) showed the reverse response.

There is less direct evidence that territory size is inversely related to predator density or intruder pressure. One of the best examples, Myers et al.'s (1979*a,b*) studies of territorial shorebirds, indicated that variation in territory size was more strongly related to intrusion pressure than to resource richness. Moreover, intrusion pressure covaried with resource richness. This implies that spatial or temporal variability in resource richness may affect territoriality indirectly through intrusion pressure by neighboring animals.

6.2.3 Optimal Territory Size and Population Dynamics

If optimal territory size (which we will term α) is inversely related to resource carrying capacity (K) and intrusion pressure (P), according to the relations implied by equation 6.12, then the rate of change of both prey and predators should depend on predator density relative to $1/\alpha$. Using the general structure outlined in Chapter 1, the rate of change of prey can be calculated according to

$$\text{if } P \leq \frac{1}{\alpha} \text{ then } \frac{dN}{dt} = rN\left(1 - \frac{N}{K}\right) - \frac{aNP}{1 + ahN} \tag{6.13}$$

$$\text{if } P \geq \frac{1}{\alpha} \text{ then } \frac{dN}{dt} = rN\left(1 - \frac{N}{K}\right) - \frac{aN}{\alpha(1 + ahN)} \tag{6.14}$$

Note that we here assume that territories never get smaller than α. The rate of change of predators should also be affected by predator density relative to the optimal territory size:

$$\text{if } P \leq \frac{1}{\alpha} \text{ then } \frac{dP}{dt} = \frac{aceNP}{1 + ahN} - (d + \theta\alpha P)\,P \tag{6.15}$$

$$\text{if } P \geq \frac{1}{\alpha} \text{ then } \frac{dP}{dt} = \frac{aceN}{\alpha(1 + ahN)} - (d + \theta\alpha P)\,P \tag{6.16}$$

As in our treatment of other behavioral phenomena, we evaluated the effect of territoriality by comparing the range of parameter combinations that are stable for the null Lotka–Volterra model and the modified model with optimal territory formation. As shown in Fig. 6.8*B*, the model with territorial predators has a substantially larger proportion of stable parameter combinations than the null model (Fig. 6.8*A*).

The explanation for this is straightforward. At low predator densities, the interaction between predators and their prey is virtually identical to that of the null model, as illustrated by their zero isoclines (Fig. 6.9). However, at high predator density, optimal territory formation leads to some predators being prevented access to resources. Hence, the predator zero isocline bends from vertical to a more horizontal orientation. Such effects on the predator zero isocline and its implications for stability have been anticipated by several previous graphical

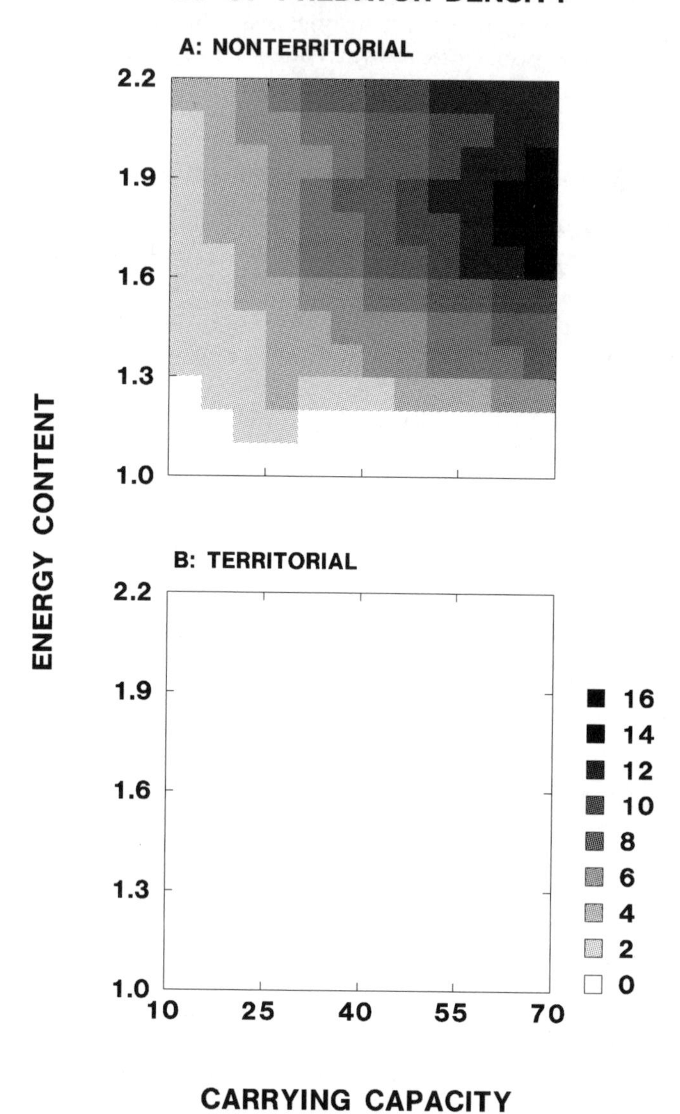

Figure 6.8. Variability in simulated predator populations in relation to energy content of prey and carrying capacity of the environment, for systems with nonterritorial predators (A), or predators defending optimal-sized territories with exclusion (B). Parameter values were as follows: $a = h = c = d = r = 1$, $\theta = 0.05$.

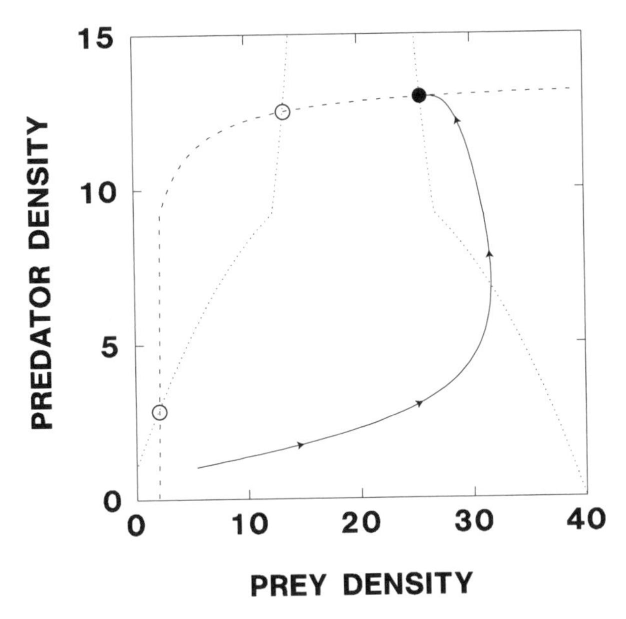

Figure 6.9. Phase plane portrait of predator density in relation to prey density for systems with territorial predators. Zero isocline of predators (dashed line) and prey (dotted line) are also shown. Nontrivial equilibria are indicated by symbols (filled = stable, open = unstable). Parameters were as follows: $a = h = c = d = r = 1$, $\theta = 0.05$, $e = 1.5$, $K = 40$.

models of predator–prey interactions, starting with the pioneering studies by Rosenzweig and MacArthur (1963).

What has not been fully appreciated, however, is that the prey zero isocline will also change with facultative changes in territory size. When predators are scarce enough to not cause any direct territorial interference, then the prey zero isocline would be hump shaped, as in the classic null model (Rosenzweig and MacArthur 1963). When $P > 1/\alpha$, however, then the prey zero isocline would be deformed upward, because nonterritorial predators are not allowed access to resources. The net result is either one or three equilibria (Fig. 6.9), which suggests that there may be a richer range of dynamical outcomes, with multiple equilibria.

Although multiple equilibria are in principle possible, we have yet to find a parameter combination that did not have a single globally stable point, that occurring at the highest prey density. This finding indicates that multiple stable equilibria, although theoretically possible, are rare in this context. It is interesting to note, however, that perturbations of prey below the stable equilibrium point can lead to long-lived transients before stability is once again restored. Circumstantial evidence from southern African populations of territorial lions and ungulates is consistent with this scenario (Smuts 1978). National park managers had harvested wildebeest to reduce their impact on plants. When harvesting was reduced, however, wildebeest continued to decline instead of immediately recovering their former density (Smuts 1978), as one would predict from the territoriality model.

6.2.4 Systematic Foraging

We have argued that under some circumstances it may be worth defending a portion of habitat to optimize rates of energy gain and that territoriality arising in this way could have important effects on trophic dynamics. Recent work has identified a fascinating reversal of this pathway: how manipulation of prey abundance by predators can be an effective means of defending a territory (Paton and Carpenter 1984; Houston et al. 1985; Gill 1988; Possingham 1989; Lucas and Waser 1989).

Assume that the territory holder splits up its territory into I cells, each of which is visited for T time units and that the forager systematically exploits this circuit of cells. By our use of the term systematically, we imply that each cell is visited as part of an iterated sequence. Further assume that the predator is able to immediately deplete each cell once visited. As a consequence, at any point in time the territory consists of a number of patches of resources at different stages of renewal (Possingham 1989). Such an abstraction might reasonably correspond to a plant–nectivore system, such as hummingbirds visiting flowers. Each flower or inflorescence is the "cell," with each visited flower being drained completely. The logistic model would be inappropriate for resource renewal under these circumstances, because renewal might well be highest when nectar has been recently depleted, although it is still reasonable to suppose an upper limit on nectar level (K). We can model the resource dynamics of such a systematic foraging system in the following manner (Possingham 1989):

$$\frac{dN_i}{dt} = r\left(1 - \frac{N_i}{K}\right) - \beta_i N_i \qquad (6.17)$$

where β_i is the probability that the forager is in cell i at any given time. Note that resource renewal is fastest when N_i is low, unlike the logistic model. This is an assumption of donor control, implying that there is negative feedback on resource rates of change but no positive feedback. Note also that we have assumed that handling time is 0 and once a cell is entered, the rate of encounter with resources is effectively infinite.

Over time, resource levels would vary across the territory. By moving systematically from cell to cell, the forager would be always using the cell with the highest resource levels (Fig. 6.10A). More importantly, an intruder wandering at random into the territory would inevitably sample a cell with lower resource levels than that available to the territory holder (Fig. 6.10B). In a sense, the territory holder behaves as if it had perfect knowledge about the state of each cell, always moving on to the best location. Such predictable differences in rates

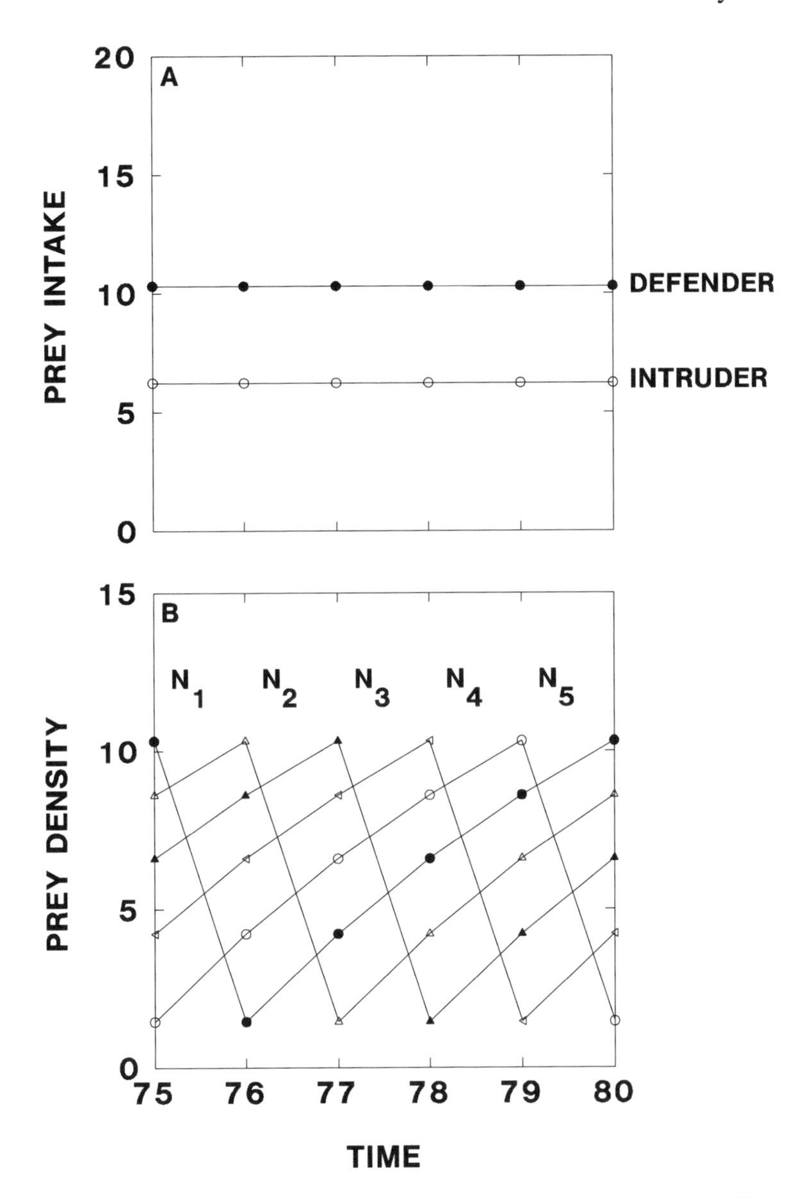

Figure 6.10. Temporal variation in food intake by a systematic territory holder (defender) and an intruder arriving at random along an array of 5 foraging sites (*A*) and variability of food levels at each foraging site (*B*), according to the model with density-independent resource renewal. Parameters were as follows: $r = 3$, $K = 20$.

of gain by intruders versus territory holders could function in the same way as agonistic displays: It would never be worth an intruder's while to steal its neighbor's resources rather than cultivating its own trapline (Lucas and Waser 1989; Possingham 1989).

Interestingly, there is a fair bit of empirical support that systematic foraging has such effects on nectarivore fitness. For example, field studies suggest that systematic foraging movements by hummingbirds can enhance rates of nectar removal relative to that obtained by random movement among flowers (Gill and Wolf 1977). More recent studies have shown that hummingbirds will adapt their visitation pattern to a fixed reward schedule imposed by researchers (Gill 1988). As predicted, territory holders obtained substantially higher reward rates than intruding neighbors. Similar systematic foraging patterns have been well documented in pied wagtails feeding on insects washed up along the banks of English streams (Davies and Houston 1984; Houston et al. 1985). In this case, food is added at a roughly constant rate, but the rate at which insects are washed away is proportional to food density. Wagtails are apparently very efficient at clearing food from the banks, so exploitation can be viewed as roughly instantaneous, at least at the densities normally encountered in nature. Hence, the resource dynamics correspond rather closely to that of hummingbirds or sunbirds exploiting nectar from flowers.

This "defense by exploitation" mechanism can be readily adapted to predict optimal territory size, where cell number is used as a surrogate measure of territory size. As additional cells are added, of course the time between exploitation events lengthens, hence the resource level per cell also increases. The mean resource density encountered by a perfectly systematic forager increases with territory size. Because there is a limit on the amount of resources per cell, however, the rate of energy gain would decelerate with increasing territory size. If one assumes that the cost of exploiting a territory increases linearly with territory size, then the diminishing energetic returns imply an optimal territory size (Fig. 6.11) (Noy-Meir 1976; Houston et al. 1985). Moreover, there are also lower and upper limits to territory size, at which an intruder has equal fitness as the territory holder. Hence, all the attributes of optimal territoriality derived from active defense can also be derived from the "defense by exploitation" mechanism.

Unfortunately, this mechanism probably cannot be generalized to other systems in which resource exploitation is constrained by rates of encounter within patches and the handling time per prey. For example, assume that resource dynamics within cells are determined by the following equation:

$$\frac{dN_i}{dt} = rN_i\left(1 - \frac{N_i}{K}\right) - \frac{a\beta_i N_i}{1 + ahN_i} \tag{6.18}$$

where β_i = the probability of visiting patch i (0 or 1 in this case) and all other parameters are as used earlier in this book. Provided that prey handling time is short and encounter rates and renewal rates are high enough, boom and bust

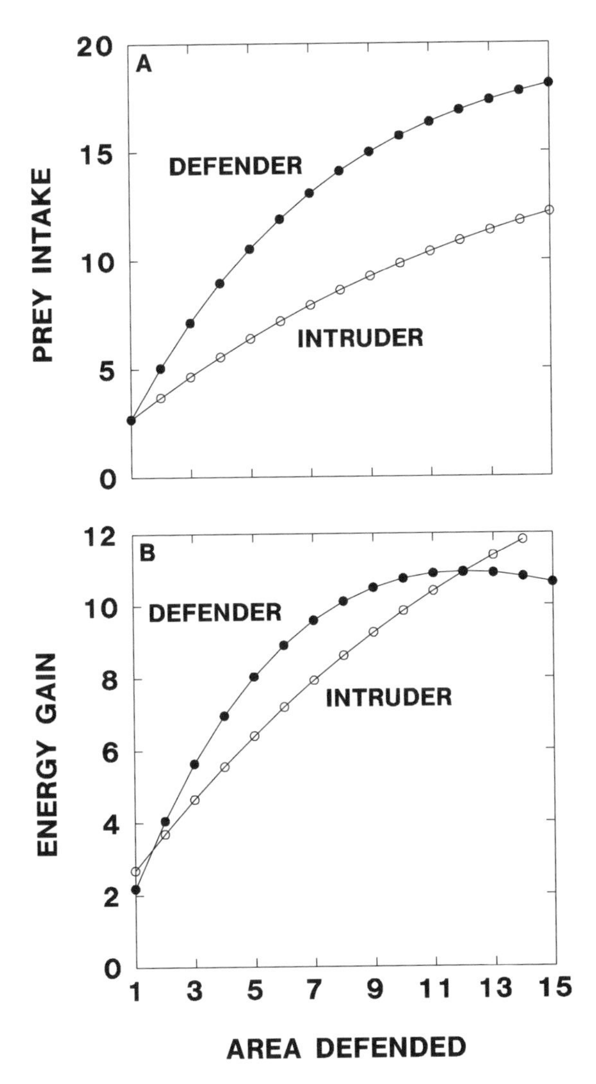

Figure 6.11. Rates of food intake (*A*) and fitness (*B*) of systematic foragers (labeled defenders) versus random intruders in relation to the number of discrete foraging sites in the systematic circuit, which directly relates to territory size. Parameters were as follows: cost = 0.5, $r = 3$, $K = 20$.

cycles of patch resource abundance would occur (Fig. 6.12*B*). As resource density within patches declines, however, the rate of consumption also falls, because encounter rates become limiting. As a result, patch depletion implies self-interference by territory holders. Intruders would therefore do at least as well as territory holders, and for most parameter combinations a good deal better than territory

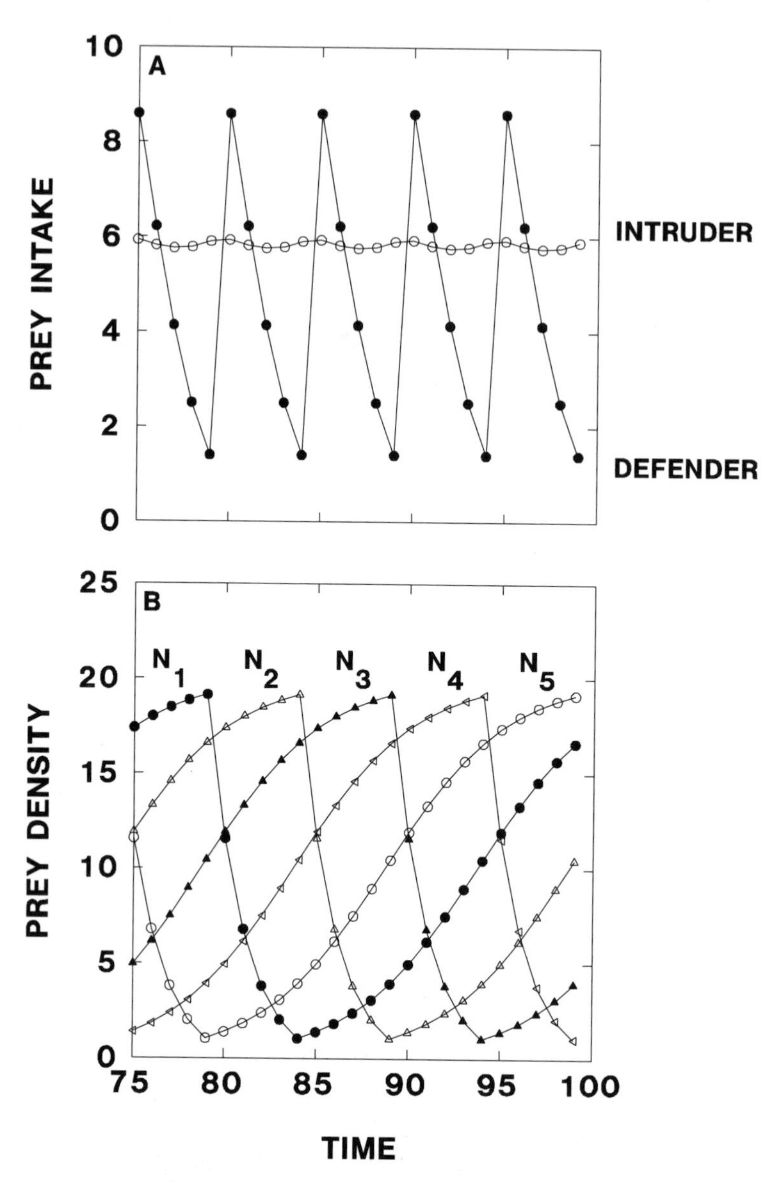

Figure 6.12. Temporal variation in food intake by a systematic territory holder (defender) and an intruder arriving at random along an array of 5 foraging sites (*A*) and variability of food levels at each foraging site (*B*), according to the model for density-dependent resource renewal. Parameters were as follows: $a = 1$, $h = 0.05$, $r = 0.3$, $K = 20$.

holders (Fig. 6.12*A*). Hence, defense by exploitation is an unlikely explanation for territory dynamics in species with constrained rates of prey consumption. Because these are also the same circumstances under which community instability can arise via trophic interactions, it suggests that systematic foraging within territories is unlikely to have an important impact on community dynamics in general.

6.2.5 Central Place Foraging and Prey Spatial Refugia

Many territorial predators have a fixed spatial frame of reference to a central place, such as a nest or burrow, from which foraging excursions are made (Andersson 1978; Schoener 1979; Orians and Pearson 1979). A central frame of reference could have interesting implications for the spatial distribution of foraging effort. Alternatively, the spatial distribution of foragers could reflect avoidance of aggressive interactions with neighboring territory holders. We first outline a Markov chain model of resource use as a function of distance from the central place and explore the effects of such spatial effects on local rates of predator search and resource use. We imbed the Markov chain model of resource use in Lotka–Volterra models to explore the implications of central place foraging vis-à-vis long-term predator–prey dynamics and competitive coexistence between prey. Finally, we consider other sources of spatial refugia in territorial systems.

Consider a hypothetical forager that cannot identify prospective prey items from the central place and instead must seek out prey during discrete bouts of food-searching behavior. Assume that the forager travels at a constant velocity along a randomly oriented vector outward from the central place, all the while searching a path of constant width on either side for suitable prey. Once encountered, single prey items are retrieved back to the central place for processing, consumption, and digestion before the next foraging bout can begin. If no prey are encountered within a maximum foraging radius from the central place, then the forager is assumed to return via the same route to the central place before initiating a new search. In this generalized foraging process, the probability that prey capture occurs at a given distance during a foraging bout is clearly conditional on the probability that no prey were successfully encountered earlier in the foraging bout.

Mathematicians often call such a pattern of system organization a Markov chain, implying that forager position is a discrete state variable, only one such state can be occupied at any point in time, and the probability of transition from one state to other states depends purely on the current state (Stewart 1994). Hence, one can model central place foraging as a Markov chain of successive search failures that is terminated by a successful prey capture.

More formally, assume a circular territory of constant radius. This circular territory can be thought of as comprising a series of concentric rings (labeled $i = 1, 2, \ldots, I$) of constant width. For convenience, we will refer to these concentric rings as distance intervals. For example, a circular territory with a radius of 100 m could be defined as comprising 100 distance intervals of 1-m width. Following standard probability theory (Scheaffer and Mendenhall 1975), the probability that a capture during a single foraging bout occurs within distance interval i can be calculated by the probability that no prey are encountered as the forager traverses distance intervals 1 to $i - 1$ multiplied by the conditional probability that at least one prey falls within the forager's search path within distance interval i. Assuming that the local spatial distribution of prey is random, these probabilities of prey encounter should follow a Poisson distribution, the parameters of which are determined by the forager's sensory radius (u), the width of each distance interval (w), and the local density of prey per unit area within distance interval i (N_i). Hence, the probability of failure to encounter any prey from the beginning of a bout until the predator reaches distance interval i can be calculated as

$$F_i = \exp\left(-uw\sum_{j=1}^{i-1} N_j\right) \tag{6.19}$$

Note that F_i is the cumulative probability within a given bout that the forager does not encounter any prey in distance intervals $i = 1$ to $i - 1$, hence the summation. The conditional probability of successful capture within distance interval i (given that the forager reaches that distance interval) is

$$S_i = 1 - \exp(-uwN_i) \tag{6.20}$$

These distance-dependent probabilities of search success can be used to calculate the expected rate of success per foraging bout and the expected time expenditures in travel and handling prey. The ratio of success rate over time expenditure will be used to estimate the overall consumption rate by a single forager (i.e., its functional response) and the risk of prey mortality due to predation in each distance interval.

The probability during any given foraging bout that a prey item is found within distance interval i is calculated as S_iF_i. By definition, the expected value of any discrete random variable is $E(Z) = \Sigma\, zg(z)$, where z is the value of the random variable and $g(z)$ is the probability density function (Scheaffer and Mendenhall 1975). The expected foraging success per foraging bout is therefore $\Sigma\, S_iF_i$, summed over the number of distance intervals ($i = 1$ to I). Similarly, the expected handling time $= \Sigma\, hS_iF_i$, where h is the time it takes to handle a single prey item. The expected time spent traveling per foraging bout can be calculated by considering the full range of potential outcomes: The forager can obtain an item in any of I distance intervals or might fail to obtain any prey at all. Multiplying the probabilities of each of these outcomes by the time it takes to travel to and from distance interval i yields the following expression for expected travel time:

$$E(\text{travel time}) = \left(\frac{2w}{v}\right)\left(\sum_{i=1}^{I} iF_iS_i + IF_I(1 - S_I)\right) \tag{6.21}$$

where v = forager travel velocity. With a bit of algebraic rearrangement, the expected travel time can be simplified to

$$E(\text{travel time}) = \left(\frac{2w}{v}\right)\sum_{i=1}^{I} F_i \tag{6.22}$$

The functional response can be calculated by dividing the expected foraging success per bout by the expected travel time plus handling time per bout (Turelli et al. 1982; Mangel and Clark 1988):

$$X = \frac{\sum_{i=1}^{I} S_iF_i}{\sum_{i=1}^{I}\left(\frac{2w}{v} + hS_i\right)F_i} \tag{6.23}$$

The number of prey deaths in distance interval i per unit time can be calculated by $X_i = XS_iF_i/\Sigma S_iF_i$, where $S_iF_i/\Sigma S_iF_i$ is the relative proportion of successful encounters occurring in distance interval i. In order to model changes in prey population density, one needs to divide the distance-specific mortality rate (X_i) by the area (A_i) of distance interval i, where $A_i = \pi w^2(2i - 1)$.

Assume that prey are initially distributed randomly, with any subsequent changes in spatial distribution strictly due to depletion by the forager. The risk of predation decreases exponentially with distance as predators first begin to forage (Fig. 6.13*A*). This leads to more rapid depletion of prey near the central place than farther away (Fig. 6.13*B*). As prey become depleted close to the central place, the forager has a lower probability of encountering a prey item early in each foraging bout and hence should tend to move farther afield with each subsequent foraging bout. Note that the model does not necessarily demand an optimal search allocation strategy on the part of the forager (Andersson 1978, 1981), but is merely a consequence of increased probability of search failure when prey are scarce. At advanced stages of depletion, predation risk is normally distributed with respect to distance (Fig. 6.13*A*).

This pattern of spatially biased resource use could have interesting effects on the functional response to changes in prey abundance and prey spatial distribution. Consumption rates should be initially high, because prey are just as common

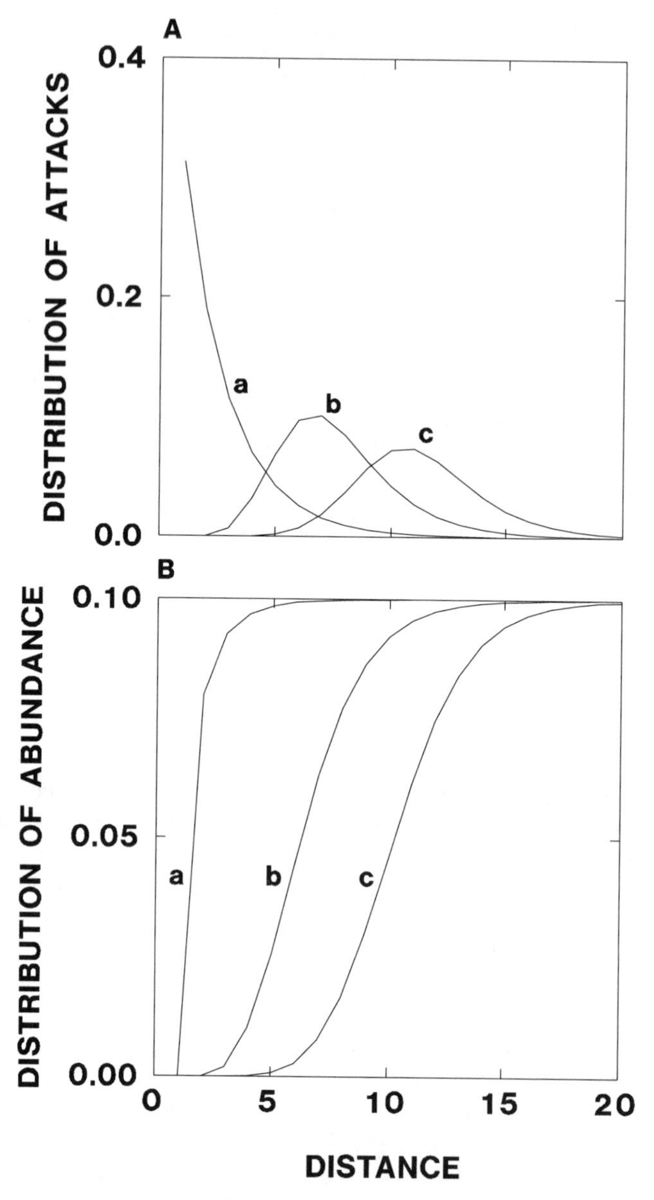

Figure 6.13. Markov chain model predictions of predation risk (*top*) and prey density (*bottom*) in relation to distance from the central place. Forager is assumed to invade a pristine habitat in which resources are homogeneously distributed at the beginning of the experiment and resources are depleted without replacement over time. Curve a depicts foraging data at the beginning of the experiment with curves b and c at later stages of resource depletion. The following parameter values were used: $u = h = 1$, $v = 10$, $I = 20$, N_i was initially set at 0.1.

close to the central place as they are farther away (Fig. 6.13*B*). However, local depletion close to the central place rapidly increases the mean distance at which food items are acquired. This should result in a steady decline in the rate of consumption, because of a considerable increase in search time, even though prey are common far away from the central place. The net result is a functional response that is concave upward at high prey densities, but concave downward at low prey densities (Fig. 6.14). Hence, spatially biased resource depletion should produce an inflection in the predator functional response, which implies positive density dependence in the per capita risk of predation mortality for prey, at least over some prey densities. Unlike most other behavioral mechanisms that produce inflections in the functional response at low prey densities (e.g., diet expansion or switching; Murdoch and Oaten 1975; Abrams 1982; Sih 1984; Fryxell and Lundberg 1994), central place foraging should produce an inflection in the functional response at moderate to high prey densities.

For comparison, the functional response of a noncentralized forager can be calculated according to a standard disc equation model ($X = uvN/(1 + uv(h + w\bar{i}/v)N))$), where the area searched per unit time equals uv, the average handling plus provisioning time equals $h + w\bar{i}/v$, and the average distance to prey equals $\bar{i}/2$. The foraging parameters u, v, h, and I in the null model have identical meanings as in the central place model. However, this alternate model carries

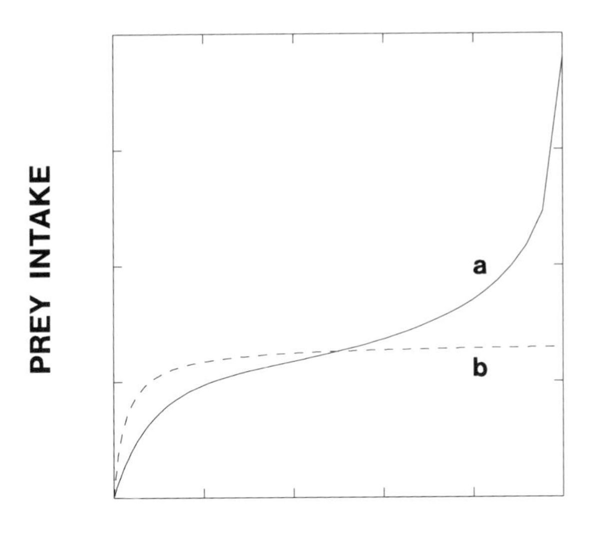

Figure 6.14. Daily rate of prey consumption in relation to mean prey density, for a hypothetical central place forager (curve a) or a random forager (curve b) following invasion into a pristine habitat. Parameter values and prey density as in Figure 6.13.

the implicit notion that the probability that the forager reaches distance i is constant. In the null case, the shape of the predator functional response is of course, always monotonically saturating (Fig. 6.14). The noncentralized forager has a lower rate of prey capture than the central place forager when prey are common, but a higher rate of capture when prey are scarce (Fig. 6.14), because the mean capture distance for the noncentralized forager is higher than that of the central place forager when prey are not yet depleted close to the central place. The converse is true when prey have been heavily depleted close to the central place. In the case of the noncentralized forager, local densities of prey are always equal.

The shape of the functional response of a central place forager is dependent on the relative magnitudes of the width of the search path (u) and the velocity (v), whereas outcomes of the disc equation are identical as long as the product uv remains constant. When a central place forager has a wide search path (u large) but low velocity, then prey depletion tends to be concentrated close to the starting point, which enhances the degree of positive density dependence by making the functional response curve more upward concave. In contrast, when the search path is narrow but velocity is high, then the inflection in the functional response curve would be relatively modest. Hence, there might be important ecological implications to variation in trade-offs between foraging efficiency versus velocity in central-place foragers.

The Markov chain model is qualitatively consistent with a number of field studies in beavers showing that the frequency of cut trees or saplings tends to decline with increasing distance from water (Hall 1960; Jenkins 1980; Pinkowski 1983; Belovsky 1984; McGinley and Whitham 1985; Basey et al. 1988). Inverse relationships between predation or activity rates and distance have been documented for other central place foragers as well (Getty 1981; Andersson 1981; Huntley et al. 1986; Greig-Smith 1987; Pickup and Chewnings 1988; Martinsen et al. 1990; Pickup 1994), and are circumstantially consistent with distance-specific patterns of seedling survival (Howe and Primack 1975) expected from distance-specific seed predators (Janzen 1970, 1971).

6.2.6 Central Place Foraging and Population Dynamics

We explored the dynamical implications of central place foraging by imbedding the Markov chain model in our standard Lotka–Volterra model of predator–prey interactions:

$$\frac{dN_i}{dt} = rN_i\left(1 - \frac{N_i}{K}\right) - PX_i \quad \text{for } i = 1 \text{ to } I \tag{6.24}$$

$$\frac{dP}{dt} = P(ceX - d) \tag{6.25}$$

where X is the functional response of a single predator, calculated according to equation 6.24. Note that prey recruitment occurs locally and the demographic response depends on local prey density, rather than prey density averaged over the entire habitat. Such neighborhood interactions are most plausible for plants (e.g., Pacala and Silander 1985, 1990), but are also conceivable for animals, provided that prey mobility is markedly less than that of their attackers. Note also that we haven't factored in the increased energy expenditure for foraging at great distance from the central place. One could conceivably handle this by making net energy gain (e) a function of distance. For simplicity, we ignore such additional costs.

Simulation results suggest that centralized foraging could have a substantial stabilizing effect on predator–prey interactions (Fig. 6.15). At many parameter combinations that are unstable for the null model (i.e., large values of K and either extremely large e or extremely small values of e), the central place model approached stability. In the portion of the parameter space in which both models were unstable, the magnitude of variation was always greater for the noncentralized foraging model than for the central place model.

It is difficult to know exactly why the central place model tends to be more stable than the null model, because the functional response depends on both the spatial distribution and mean density of prey, but there are at least two possibilities. First, the inflection in the functional response of the central place forager (Fig. 6.14) implies that prey mortality should be positively density dependent over some ranges of prey density, unlike the negative density dependence of the characteristic of the disc equation (Murdoch and Oaten 1975). It is well known that such positive density dependence can be stabilizing, at least under some circumstances (Murdoch and Oaten 1975; Tansky 1978; Abrams 1982; Sih 1984; Fryxell and Lundberg 1994).

A second, and more likely, explanation for the difference in stability properties relates to the lower rate of consumption by the central place forager compared to the noncentralized forager when resources are depleted close to the central place. This condition implies that the central place forager is less efficient at exploiting prey when prey are scarce than is the noncentralized forager, and foraging efficiency at low prey density is inversely associated with local stability (Murdoch and Oaten 1975). In either case, these simulations strongly suggest that central place foraging could potentially be important in stabilizing long-term dynamics.

At least two other processes would lead to foragers being found most of the time near the center of their territories: avoidance of aggressive conspecifics or avoidance of other predators. Territoriality usually involves agonistic interactions between neighbors that a prudent territory-holder might want to avoid. For example, aggression among spotted hyenas is most often recorded near territory boundaries (Hofer and East 1993*a,b*). Long-term demographic data for wolves suggest that mortality due to territorial aggression is pronounced near territorial boundaries (Mech 1994). In the case of wolves, territorial boundaries are often identified

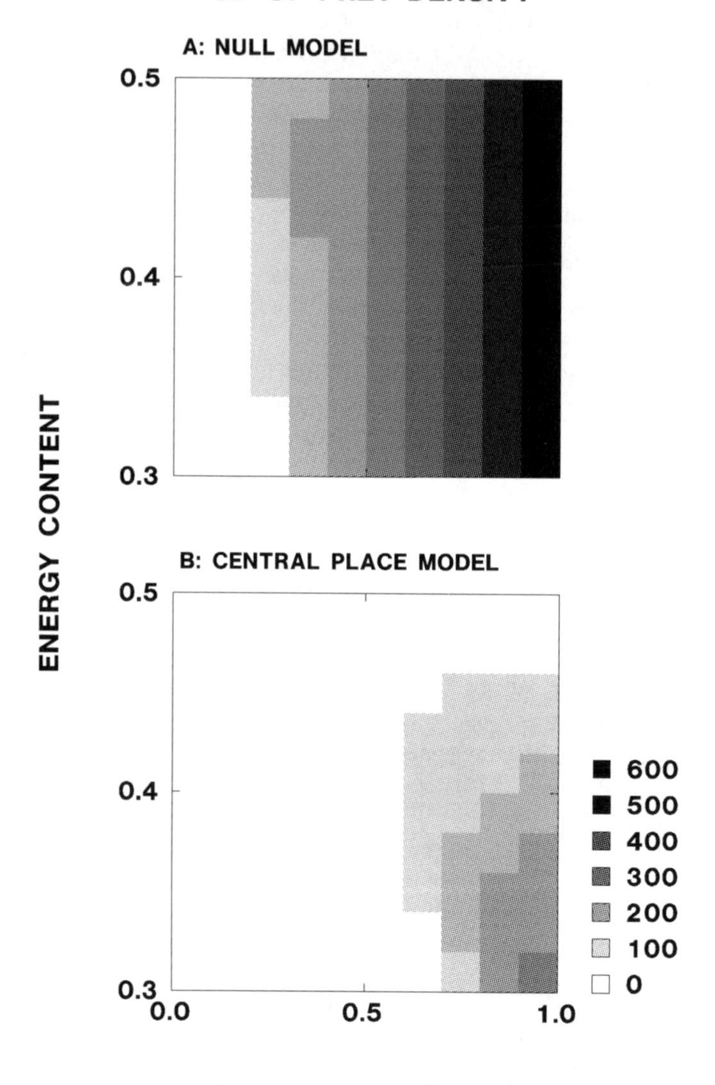

Figure 6.15. Variability over time in prey densities averaged over the entire foraging radius and averaged over the last 200 time steps of 600 time step simulations, using the central place model (*top*) and the random predator model (*bottom*) over identical parameter ranges ($u = v = 10$, $r = h = c = 1$, $I = 20$, and $d = 0.2$, other parameters as indicated along axes of figure).

by high densities of scent marks, which when detected by a passing wolf have the dual effect of stimulating more scent marking and an aversion reaction. Lewis and Murray (1993) developed an elegant model of this complex behavioral process, incorporating diffusive movements away from the central place to acquire prey but convective movements toward the organizing center in response to scent marks left by neighbors. Trade-offs between the diffusion and convection gradients produce a heterogeneous spatial distribution for both predators and prey that is analogous to that obtained from the Markov chain model of central place foraging, with predators most often found close to the territory center whereas prey are most often found at the boundary between territories (Lewis and Murray 1993). Interestingly, radio-marked deer have been shown to congregate in the interstices between wolf territories, as predicted by the convection–diffusion model and the Markov chain model (Hoskinson and Mech 1976).

Taylor and Pekins (1991) constructed several alternate models of wolf–deer interactions based on core versus boundary zones with different rates of visitation by wolves. They found that boundary avoidance enhanced stability if there is substantial dispersal by deer between core and "refuge" zones, particularly when such deer dispersal is linked to the probability of wolf attack rather than as a response to escalating deer density. However, if little dispersal occurs, if dispersal probability of deer is positively density dependent, or rates of prey search in boundary areas by wolves is inversely related to deer abundance in core areas, then system stability would be diminished. Some of this confusion in results might be traced to the rather simple nature of Taylor and Pekins's models: core versus boundary regions, linear functional responses by predators and no density-dependent response in prey recruitment.

Predation risk could also influence activity patterns around the central place. Perhaps the most elegant depiction of such a process dates from Lotka (1932). Assume that predators randomly approach prey at a given velocity, and that prey flee from detected predators back to a central refuge, and that prey have a fixed detection radius around them. If predators chase fleeing prey as they run for the central place, then the risk of predation increases exponentially with distance from the central place (Taylor 1988). A prudent forager that balanced energy gains against the risk of predation would search more intensively near the central place than one might expect from the Markov chain model. If anything, these features would reinforce spatial refugia for prey far from the territory center.

The similarity of predictions arising from the foraging and risk avoidance models suggests that many features of central place foraging tend to lead to concentration of predator activity near the central place and concentration of prey activity near territory margins, which should tend to have a stabilizing influence on trophic interactions.

6.3 SUMMARY

We examine social interactions among predators and their effects on trophic interactions. Several different forms of social interactions are considered, including direct interference, territoriality to optimize energetic gains relative to costs, systematic foraging to discourage territorial intrusion, and central place foraging. All of these forms of social behavior are strongly stabilizing, indeed more consistently stabilizing than other behaviors looked at in previous chapters.

Direct interference arises out of lost opportunities for prey search whenever a forager encounters a conspecific. As a direct consequence of mass action principles, encounters will be most frequent when foragers are most actively searching and when foragers are common. As a result, rates of prey intake should rapidly drop with increased forager density, a pattern documented in several species. Social status can only erode this stabilizing effect, because it inevitably means that some individuals are invulnerable to interference. Despotic spatial distributions according to social status probably erode the stabilizing effect of interference to a lesser degree.

There are many ways to approach territoriality, perhaps because there are many functions and ecological correlates of territoriality. It is possible to predict energetically optimal territory formation on the basis of augmentation of prey density *at equilibrium*. Escalating costs and the decreased marginal value of territories beyond the optimal size ensure set an upper bound on territory size. Optimal territory size should vary positively with resource carrying capacity, but negatively with predator population density. Optimal territory formation is strongly stabilizing, because it protects resources from overly intense predation rates and because it imposes limits on predator densities.

Systematic foraging can arise as a nonaggressive means of territorial defense. It only works when resource renewal is not logistic, however, so it is most likely to occur in donor-controlled systems. In interactive predator–prey systems, systematic foraging would rarely prove to be an evolutionarily stable strategy.

Central place foraging leads to interesting effects on rates of resource utilization at varying distance from the central place. Mechanistic considerations alone dictate that depletion will initially be concentrated close by, but will eventually shift some distance away from the central place. Skewed spatial distribution of resources over time causes an increased time expenditure in foraging, which reduces feeding efficiency when prey become scarce. This reduction in feeding efficiency due to skewed resource distribution in space has a stabilizing effect on trophic interactions.

7 Epilogue

At the beginning of this book we asked whether adaptive behavior might explain the apparent persistence, if not stability, of real communities. We hope that by now even the most skeptical reader is convinced that adaptive forms of behavior might play a major influence on community dynamics. A growing body of work suggests that adaptive behavior has neither a routinely stabilizing influence, nor a routinely destabilizing influence. It all depends on the form of behavior, the environmental structure, and the set of environmental parameters that apply in a given situation.

Take, for example, the effects of adaptive diet selection by predators (Chapter 2). If natural selection favors foragers that select a diet that balances the mix of nutrients and/or energy, then this "adaptive" behavior is unlikely to lend much of a stabilizing influence. On the other hand, natural selection for maximization of rates of energy gain can lend a stabilizing influence on predator–prey dynamics, provided that a number of conditions apply. Selection among prey that when superabundant are more than adequate to sustain predators is again unlikely to have a stabilizing influence. On the other hand, if low-ranking prey *are* just sufficient to sustain predators, then adaptive diet choice could lend a stabilizing influence. In all cases, diet selection is more likely to lend a stabilizing influence at the population level when there is pronounced behavioral variation present in the forager population or when foragers are highly variable in their individual diet choices.

Nonetheless, we would now argue that some forms of behavior are much more likely to have important community implications than others. For example, the work covered in Chapter 5 suggests that age- or size-dependent feeding selectivity has surprisingly little effect at the population level. This conclusion is somewhat disturbing, because it implies that dozens of published field tests of size selectivity probably have little dynamical application. On the other hand, almost all of the forms of interference and territoriality that we examined in Chapter 6 have strong stabilizing effects at the population level, yet this field has attracted little recent interest.

These conclusions suggest a real danger in ecology. It is a complicated world out there, and without theoretical guidance it is very difficult to know which areas of behavioral research will prove most illuminating at other levels of organization. We see disappointingly little evidence of sustained interest by behavioral ecologists in the wider implications of behavioral patterns. For example, few empirical studies of habitat choice have interpreted their results in terms of population effects (Werner et al. 1983*a,b;* Brown et al. 1992b, 1994; Kotler 1992), even though it has been clear for quite some time that there are profound population implications.

There is little doubt that this schism stems from a variety of causes. It is difficult, of course, for any single scientist to keep pace with the rapidly growing literature even in his or her own narrow speciality. So, it is perhaps to be expected that the pressures of research specialization tend to mitigate against awareness of complementary work in other disciplines. With the growth of collaborative research programs of a long-term nature, the necessity of research specialization is less compelling. We suspect, however, that there may be more important sources of apathy.

Like any other area of research, behavioral ecology is subject to fads. Our subjective impression is that foraging questions are often regarded as being almost passé, perhaps as a response to repeated challenges to the optimization principle itself (Gould and Lewontin 1979). More likely, this situation reflects a practical recognition that there have already been many well-controlled tests of optimal foraging theory (e.g., Stephens and Krebs 1986). Many of the most frequently cited tests of optimal foraging theory have used unrealisticly simplistic protocols. It is not entirely clear whether the kinds of foraging decisions made by birds trained to feed upon mealworms on a conveyor belt inform us a great deal about decisions made in the field. The very success of the first generation of well-controlled, albeit unrealistic, lab experiments leaves an impression that the crucial experiments have already been done.

Attempts to extend predictions of foraging theory to the field often suffer due to a lack of crucial parameters or inherent difficulty in measuring state variables and behavioral responses. It is not a trivial matter to evaluate the energetic profitability of dozens of potential prey items. Having done so, it is perhaps even more challenging to measure variation in prey availability at scales relevant to the forager across different locations for a comparative field study. At typical population densities, it is also no trivial exercise to measure patterns of habitat use or diet selection in free-living animals. It is unlikely that one would undertake such a research program unless there was a more compelling reason. If the logical consequences of adaptive behavior are shown to have wider significance, then this recognition might encourage further effort to perform field-based tests of foraging theory.

On the theoretical front, there has been much preoccupation with developing minor variants of existing behavioral models. For example, numerous theoretical papers are based on the basic contingency model of diet selection. Most of these models make similar qualitative predictions, hence they are unlikely to be discriminated on the basis of noisy experimental data.

From the point of view of understanding community dynamics, however, preoccupation with behavioral detail deflects attention away from more pressing concerns, such as understanding the effects of qualitative features common to all reasonable behavioral models. For example, virtually the same stabilizing effect of diet selection would occur for any model in which foragers become less selective as preferred prey become less common, particularly if there is a lot of intrinsic variability in the behavioral response measured at the population level. The specifics of the behavioral model itself are therefore essentially unimportant at the population level. What we need is more theory relating to sources of variation among individuals!

By the same token, population ecologists rarely even discuss the effect of realistic patterns of behavior on population phenomena. For example, there is a rich theoretical literature dealing with metapopulation dynamics, yet almost all of these works assume a constant probability of dispersal from one site to the next. As we have shown in Chapter 4, adaptive dispersal leads to a different set of predictions than does constant dispersal, just as constant recruitment generates different population dynamics than does density-dependent recruitment.

It is particularly surprising that population ecologists pay so little attention to behavioral ecology given that the latter discipline has much stronger empirical support. We had initially hoped to find a rich empirical base of population data to illustrate this book, to match the substantial literature in behavioral ecology. We were dismayed to learn how little behavior there is in most field studies of population ecology.

This situation could be due to the fact that hypotheses are intrinsically harder to test at higher levels of organization. A field test of competition theory requires much greater investment in time and effort than a mensurative experiment on optimal foraging. Not only are the protocols more difficult, but the predictions are often murky in long-term experiments on ecological communities.

A more insidious reason, however, is that many empirical population ecologists seem to have a fundamental distrust of quantitative theory. Population ecology has a rich theoretical tradition, but very little of this body of theory forms the basis of field studies. By and large, field ecologists often tend to test verbal theories or models. The bulk of population studies still involve interpretations of population time-series, evaluation of density-dependent relationships among demographic variables and abundance, or simple "experiments" involving predator removal, nutrient enrichment, and so on. Such studies are rarely based on quantitative theory of any sort, let alone theory based on adaptive behavioral responses. Until the schism between population theory and field ecology is repaired there is little chance of testing the community implications of behavioral models.

Current trends in population ecology are scarcely reassuring on this score. It is growing harder and harder to secure funding for the kinds of long-term studies needed to test population phenomena such as predator–prey dynamics or competition. Much of the available funding has been directed toward conservation biology, often focusing on the immediate demographic and genetic risks to small populations without considering the biological processes that generate small populations in the first place (Caughley and Gunn 1996). Until there is a fundamental shift in research perspective, it is hard to believe that conservation biology will develop the potential we so vitally need in a changing world.

This brings us logically to an assessment of fruitful areas for future research. On the theoretical front, several areas would be worthy of deeper consideration. First, all of the behavioral models we have discussed are framed in terms of immediate fitness. Behavioral ecologists have begun to appreciate the full implications of state-dependent decision making, particularly in the context of short-term decision making on lifetime reproductive success (McFarland and Houston 1981; McNamara and Houston 1982, 1986; Mangel and Clark 1986, 1988). We have avoided this elegant and rapidly growing body of work for several reasons. One needs to walk before one runs, and we have tried to provide a useful introduction to a difficult interdisciplinary field rather than a definitive treatise for those at the cutting edge of research. State-dependent decision making necessitates additional wrinkles that substantially increase model complexity. Because individuals within a given population by definition differ, full-blown state-dependence requires individually based models. Basing decisions on expectations of lifetime reproductive success similarly requires age or stage structure that requires extensive bookkeeping. Calculating optimal decisions often involves iterative computer algorithms that are sometimes costly in terms of time. These additional features of state-dependent behavior complicate population modeling sufficiently to require separate treatment. We would enthusiastically endorse such a research program, given the impressive predictive success of state-dependent models (Mangel and Clark 1988).

In the course of writing this book, it became obvious that some areas are much more richly developed than others. The chapter on predator interference and territoriality specifically comes to mind. A multitude of different forms of social interaction need to be explored in the context of community dynamics. For example, niche partitioning occurs in many species, frequently on the basis of gender, age, or social status. We have already demonstrated potential for unusual dynamics among competitors linked by partial overlap in resource use. What happens when these competitors share a common gene pool?

There is much room for further consideration of frequency-dependent trade-offs in ecological fitness of solitary individuals versus those in groups. Although we did consider to some degree the effect of varying social status within a consumer population, we gave little consideration to the explicit mechanism by which social status is actually achieved. In many species, size is synonymous with social status, by virtue of its effect on agonistic or aggressive activities. Recent studies suggest that size structure in consumer populations can have interesting dynamical effects. We would be surprised if size-structured dominance did not have interesting implications for community dynamics, due to cohort-induced time lags.

The spatial dynamics of predators and prey clearly requires further work—particularly when movements are costly in terms of other life history traits or when the landscape itself is complex. For example, it clearly makes sense for potential prey to avoid predators, all other things being equal (Chapter 4). But there are many ways by which an organism might balance current risk against long-term prospects for reproductive success, presenting a wide range of migratory habitat use strategies that evolutionary ecologists are only beginning to understand (Mangel and Clark 1988; Dingle 1994). Like most metapopulation theory, our spatially structured models assumed simple Cartesian coordinates in a landscape without topographic detail. Real landscapes, unlike their silicon counterparts, are rich in topographic detail, which one might expect to influence patterns of resource use. Like most other biologists, we assumed unbiased movement patterns across the landscape. But real animals probably do not exhibit unbiased movements in space, particularly when resource distributions are contagiously distributed. More realistic models of diffusive movements among spatial locations offer exciting research opportunities.

We also foresee much potential for explicit modeling of genetics, behavior, and population dynamics. The study of the genetic basis of both ecological interactions and behavior is admittedly in its infancy (Boake 1994; Real 1994). Nonetheless, much progress has been made in quantifying the magnitude of genetic versus nongenetic contributions to behavioral variability (Boake 1994). Just as importantly, it is clear that specific genetic systems can have important effects on the dynamics of pathogen–host (Antonovics 1994), parasitoid–host (Doebelli 1996), or predator–prey systems (Saloniemi 1993; Fryxell 1997). If the last two decades have seen a rapid expansion in our theoretical understanding of behavior on community dynamics, we hope that the next decade witnesses similar links between population genetics and ecology.

On the empirical front, there are huge gaps in our understanding of the linkage between behavior and population ecology, particularly regarding behavioral responses to changing population densities. Most of the ecologically important interactions we have discussed in this book relate to compensatory responses to resource scarcity or risk of predation. Such responses are well documented in the lab but poorly described under field conditions.

One useful starting point might be to revisit the neglected concept of the functional response. Virtually all of the behavioral models that we have discussed have demonstrable effects on the rates of consumption. By setting experimental conditions so as to favor alternate decisions, it is possible to test predicted effects on functional and numerical responses (Murdoch et al. 1975; Ranta and Nuutinen 1985; Colton 1987; Lundberg and Danell 1990; Mitchell and Brown 1990; Fryxell and Doucet 1993). Of course, this testing will require measurement of functional responses in experiments with multiple prey species, size classes, or spatial arrangements of prey. As functional and numerical responses are the cornerstones of adaptive behavioral models, tests of effects on the functional response represent the simplest rigorous tests for rejecting behavioral hypotheses.

Although behavioral tests are certainly more feasible than testing predictions at the population level, there is also need for comparative behavioral studies that test for behavioral changes across populations over time or space. Such mensurative data would seldom prove sufficient by themselves for hypothesis rejection, but can be compelling when used in conjunction with more detailed behavioral experiments.

Long-term population studies of any sort are, of course, in extremely short supply, so any additional data would be highly welcome. Time-series data on single species, however, are of rather limited utility. There is much exciting contemporary work being conducted on the analysis of time-series, such as attractor reconstruction and stochastic nonlinear forecasting. Nonetheless, it is unlikely that such work can ever yield much more than a suggestion of the ecological mechanisms at play. It is high time to see field population ecologists address multiple trophic levels. Long-term studies of entire guilds or consumer resource assemblages are all too rare.

Above all, we would endorse an interdisciplinary approach to community dynamics. Few ecologists have the breadth of experience and interests to support functional research programs at both the behavioral and population level. The time and effort required to adequately test behaviorally complex interactions also mitigate against single-person studies. Logical candidates for such an approach should be sites of long-term studies, with a rich database at the population level to guide the requisite behavioral and population experimentation.

References

Abrahams, M. V. 1986. Patch choice under perceptual constraints: a cause for departures from an ideal free distribution. *Behavioral Ecology and Sociobiology* 19: 409–415.

Abrams, P. A. 1982. Functional responses of optimal foragers. *American Naturalist* 120: 382–390.

——— 1983. Life-history strategies of optimal foragers. *Theoretical Population Biology* 24: 22–38.

——— 1984. Foraging time optimization and interactions in food webs. *American Naturalist* 124: 80–96.

——— 1986. Adaptive responses of predators to prey and prey to predators: the failure of the arms-race analogy. *Evolution* 40: 1229–1247.

——— 1987*a*. The functional responses of adaptive consumers of two resources. *Theoretical Population Biology* 32: 262–288.

——— 1987*b*. Indirect interactions between species that share a common predator: varieties of indirect effects, pp. 38–54, in W. C. Kerfoot and A. Sih, eds., *Predation: Direct and Indirect Impacts on Aquatic Communities,* University of New England Press, Dartmouth, N.H., U.S.A.

——— 1989. Decreasing functional responses as a result of adaptive consumer behavior. *Evolutionary Ecology* 3: 95–114.

——— 1990*a*. The evolution of anti-predator traits in prey in response to evolutionary change in predators. *Oikos* 59: 147–156.

——— 1990*b*. The effects of adaptive behavior on the type-2 functional response. *Ecology* 71: 877–885.

——— 1991. Life history and the relationship between food availability and foraging effort. *Ecology* 72: 1242–1252.

——— 1993. Effect of increased productivity on the abundance of trophic levels. *American Naturalist* 141: 351–371.

——— 1994*a*. Should prey overestimate the risk of predation? *American Naturalist* 144: 317–328.

——— 1994*b*. Evolutionary stable growth rates in size-structured populations under size-related competition. *Theoretical Population Biology* 46: 78–95.

Abrams, P., and H. Matsuda. 1993. Effects of adaptive predatory and anti-predator behaviour in a two-prey-one-predator system. *Evolutionary Ecology* 7: 312–326.

Abrams, P. A. and J. Roth. 1994*a*. The responses of unstable food chains to enrichment. *Evolutionary Ecology* 8: 150–171.

Abrams, P. A. and J. Roth. 1994*b*. The effects of enrichment of three-species food chains with non-linear functional responses. *Ecology* 75: 1118–1130.

Abrams, P. A. and L. Shen. 1989. Population dynamics of systems with consumers that maintain a constant ratio of intake rates of two resources. *Theoretical Population Biology* 35: 51–89.

Abramsky, Z., M. L. Rosenzweig, B. Pinshow, J. S. Brown, B. Kotler, and W. A. Mitchell. 1990. Habitat selection: an experimental field test with two gerbil species. *Ecology* 71: 2358–2369.

Abramsky, Z., M. L. Rosenzweig, and B. Pinshow. 1991. The shape of a gerbil isocline measured using principles of optimal habitat selection. *Ecology* 72: 329–340.

Adler, F. R. and C. D. Harvell. 1990. Inducible defenses, phenotypic variability and biotic environments. *Trends in Ecology and Evolution* 5: 407–410.

Adler, F. R. and R. Karban. 1994. Defended fortresses or moving targets? Another model of inducible defenses inspired by military metaphors. *American Naturalist* 144: 813–832.

Akre, B. G. and D. M. Johnson. 1979. Switching and sigmoid functional response curves by damselfly naiads with alternative prey available. *Journal of Animal Ecology* 48: 703–720.

Andersson, M. 1978. Optimal foraging area: size and allocation of search effort. *Theoretical Population Biology* 13: 397–409.

Andersson, M. 1981. Central place foraging in the whinchat, *Saxicola rubetra. Ecology* 62: 538–544.

Antonovics, J. 1994. The interplay of numerical and gene-frequency dynamics in host-pathogen systems. Pp. 129–145 in L. A. Real, ed., Ecological genetics, Princeton University Press, Princeton, N.J., USA.

Armstrong, R. A. and R. McGehee. 1980. Competitive exclusion. *American Naturalist* 115: 151–170.

Åström, M. and P. Lundberg. 1994. Plant defence and stochastic risk of herbivory. *Evolutionary Ecology* 8: 288–298.

Åström, M., P. Lundberg, and K. Danell. 1990. Partial prey consumption by browsers: trees as patches. *Journal of Animal Ecology* 59: 287–300.

Augner, M. 1995. Plant-plant interactions and the evolution of defences against herbivores, Ph.D. thesis, Lund University, Lund, Sweden.

Bach, C. E. 1994. Effects of herbivory and genotype on growth and survivorship of sand-dune willow (*Salix cordata*). *Ecological Entomology* 19: 303–309.

Basey, J. M., S. Jenkins, and P. E. Busher. 1988. Optimal central-place foraging by beavers: tree-size selection in relation to defense chemicals of quaking aspen. *Oecologia,* (Berl.) 76: 278–282.

Bayliss, P. 1985. The population dynamics of red and western grey kangaroos in arid New South Wales, Australia. II. The numerical response function. *Journal of Animal Ecology* 54: 127–135.

Beckett, P. H. T. and R. Webster. 1971. Soil variability: a review. *Soils Fertility* 34: 1–15.

Beddington, J. R. 1975. Mutual interference between parasites or predators and its effect on searching efficiency. *Journal of Animal Ecology* 44: 331–340.

Begon, M. and M. Mortimer. 1986. *Population Ecology—A Unified Study of Animals and Plants.* Blackwell Scientific Publications, Oxford.

Begon, M., J. L. Harper, and C. R. Townsend. 1996. *Ecology,* 3rd edition. Blackwell, Oxford.

Bell, G., M. J. Lechowicz, A. Appenzeller, M. Chandler, E. DeBlois, L. Jackson, B. Mackenzie, R. Preziosi, M. Schallenberg, and N. Tinker. 1993. The spatial structure of the physical environment. *Oecologia* (Berl.) 96: 114–121.

Belovsky, G. E. 1978. Diet optimization in a generalist herbivore: the moose. *Theoretical Population Biology* 14: 105–134.

Belovsky, G. E. 1984. Summer diet optimization by beaver. *American Midland Naturalist* 111: 209–222.

Belovsky, G. E. 1986. Generalist herbivore foraging and its role in competitive interactions. *American Zoologist* 26: 51–69.

Bence, J. R. and W. W. Murdoch. 1986. Prey size selection by the mosquitofish: relation to optimal diet theory. *Ecology* 67: 324–336.

Bernays, E. A., K. L. Bright, N. Gonzalez, and J. Angel. 1994. Dietary mixing in a generalist herbivore: tests of two hypotheses. *Ecology* 75: 1997–2006.

Bernstein, C. 1984. Prey and predator emigration responses in the acarine system *Tetranychus urticae–Phytoseiulus persimilis. Oecologia* (Berl.) 61: 134–142.

Bernstein, C., A. Kacelnik, and J. R. Krebs. 1988. Individual decisions and the distribution of predators in a patchy environment. *Journal of Animal Ecology* 57: 1007–1026.

Bernstein, C., A. Kacelnik, and J. R. Krebs. 1991. Individual decisions and the distribution of predators in a patchy environment. II. The influence of travel costs and structure of the environment. *Journal of Animal Ecology* 60: 205–225.

Bjorndal, K. A. 1991. Diet mixing: nonadditive interactions of diet items in an omnivorous freshwater turtle. *Ecology* 72: 1234–1241.

Boake, C. R. B. 1994. Quantitative genetic studies of behavioral evolution. University of Chicago Press, Chicago, IL, USA.

Bobek, B. 1977. Summer food as the factor limiting roe deer population size. *Nature* 268: 47–49.

Brown, D. G. 1988. The cost of plant defense: an experimental analysis with inducible proteinase inhibitors in tomato. *Oecologia* (Berl.) 76: 467–470.

Brown, J. S. 1986. Coexistence on a resource whose abundance varies spatially. Ph.D. Dissertation, Department of Ecology and Evolutionary Biology, University of Arizona.

——— 1988. Patch use as an indicator of habitat preference, predation risk, and competition. *Behavioural Ecology and Sociobiology* 22: 37–47.

——— 1990. Habitat selection as an evolutionary game. *Evolution* 44: 732–746.

——— 1992. Patch use under predation risk: I. Models and predictions. *Annales Zoologici Fennici* 29: 301–309.

——— 1996. Coevolution and community organization in three habitats. *Oikos* 75: 193–206.

Brown, J. S. and P. U. Alkon. 1990. Testing values of crested porcupine habitats by experimental food patches. *Oecologia (Berl.)* 83: 512–518.

Brown, J. S. and W. A. Mitchell. 1989. Diet choice on depletable resources. *Oikos* 54: 33–43.

Brown, J. S. and N. B. Pavlovic. 1992. Evolution in heterogeneous environments: effects of migration on habitat specialization. *Evolutionary Ecology* 6: 360–382.

Brown, J. S. and T. L. Vincent. 1992. Organization of predator-prey communities as an evolutionary game. *Evolution* 46: 1269–1283.

Brown, J. S., Y. Arel, Z. Abramsky, and B. P. Kotler. 1992a. Patch use by gerbils (*Gerbillus allenbyi*) in sandy and rocky habitats. *Journal of Mammalogy* 73: 821–829.

Brown, J. S., R. A. Morgan, and B. D. Dow. 1992b. Patch use under predation risk: II. A test with fox squirrels, *Sciurus niger. Annales Zoologici Fennici* 29: 311–318.

Brown, J. S., B. P. Kotler, and W. A. Mitchell. 1994. Foraging theory, patch use, and the structure of a Negev desert granivore community. *Ecology* 75: 2286–2300.

Burrows, M. T. and R. N. Hughes. 1991. Variation in foraging behaviour among individuals and populations of dogwhelks, *Nucella lapillus:* natural constraints on energy intake. *Journal of Animal Ecology* 60: 497–514.

Butler, S. M. and J. R. Bence. 1984. A diet model for planktivores that follow density-independent rules for prey selection. *Ecology* 685: 1885–1894.

Caswell, H. 1989. *Matrix Population Models.* Sinauer, Sutherland, MA.

Caswell, H. and R. J. Etter. 1993. Ecological interactions in patchy environments: from patch-occupancy models to cellular automata. Pp. 93–109 *in* S. A. Levin, T. M. Powell, and J. H. Steele, eds., *Patch Dynamics*. Springer-Verlag, Berlin.

Caswell, H. and A. M. John. 1992. From the individual to the population in demographic models. Pp. 36–61, *in* D. L. DeAngelis and L. J. Gross, eds., *Individual Based Models and Approaches in Ecology—Populations Communities and Ecosystems*. Chapman & Hall, New York.

Caughley, G. 1976. Wildlife management and the dynamics of ungulate populations. Pp. 183–246 *in* T. H. Coaker, ed., *Applied Biology* Academic Press, London.

Caughley, G. and A. Gunn. 1996. *Conservation Biology in Theory and Practice*. Blackwell, Oxford, UK.

Cayford, J. T. and J. D. Goss-Custard. 1990. Seasonal changes in the size selection of mussels, *Mytilus edulis,* by oystercatchers, *Haematopus ostralegus:* an optimality approach. *Animal Behaviour* 40: 609–624.

Charlesworth, B. 1980. *Evolution in Age-Structured Populations*. Cambridge University Press, Cambridge.

Charnov, E. L. 1976*a*. Optimal foraging: attack strategy of a mantid. *American Naturalist* 110:141–51.

Charnov, E. L. 1976*b*. Optimal foraging, the marginal value theorem. *Theoretical Population Biology* 9: 129–136.

Clark, C. W. and C. D. Harvell. 1992. Inducible defenses and the allocation of resources: a minimal model. *American Naturalist* 139: 521–539.

Clutton-Brock, T. H., O. F. Price, S. D. Albon, and P. A. Jewell. 1991. Persistent instability and population regulation in Soay sheep. *Journal of Animal Ecology* 60: 593–608.

Colton, T. F. 1987. Extending functional response models to include a second prey type: an experimental test. *Ecology* 68: 900–912.

Comins, H. N. and D. W. E. Blatt. 1974. Prey–predator models in spatially heterogeneous environments. *Journal of Theoretical Biology* 48: 75–83.

Comins, H. N. and M. P. Hassell. 1976. Predation in multi-prey communities. *Journal of Theoretical Biology* 62: 93–114.

Comins, H. N., M. P. Hassell, and R. M. May. 1992. The spatial dynamics of host-parasitoid systems. *Journal of Animal Ecology* 61: 735–748.

Connell, J. H. 1973. Population ecology of reef building corals. Pp. 205–245, *in* O. A. Jones and R. Endean, eds., *Biology and Geology of Coral Reefs,* Vol. 2. Academic Press, New York.

Cook, R. M. and B. J. Cockrell. 1978. Predator ingestion rate and its bearing on feeding time and the theory of optimal diets. *Journal of Animal Ecology* 47: 529–547.

Cowie, R. J. 1977. Optimal foraging in great tits (*Parus major*). *Nature* 268: 137–139.

Crawley, M. J. 1983. *Herbivory*. Blackwell, Oxford.

Crowley, P. H. 1981. Dispersal and the stability of predator–prey interactions. *American Naturalist* 118: 673–701.

Cuddington, K. M. and E. McCauley. 1994. Food-dependent aggregation and mobility of the waterfleas *Ceriodaphnia dubia* and *Daphnia pulex. Canadian Journal of Zoology* 72: 1217–1226.

Curio, E. 1976. *The Ethology of Predation*. Springer-Verlag, Berlin.

Davies, N. B. and A. I. Houston. 1984. Territory economics. Pp. 148–169. *In* J. R. Krebs and N. B. Davies, eds., *Behavioural Ecology*. Blackwell, Oxford.

DeAngelis, D. L. and L. J. Gross. eds. 1992. *Individual-Based Models and Approaches in Ecology*. Chapman & Hall, New York.

DeAngelis, D. L. and K. A. Rose. 1992. Which individual-based approach is most appropriate for a given problem?, Pp. 67–82, *in* D. L. DeAngelis and L. J. Gross, eds., Individual-Based Models and Approaches in Ecology—Populations, Communities and Ecosystems. Chapman & Hall, New York.

DeAngelis, D. L., R. A. Goldstein, and R. V. O'Neill. 1975. A model for trophic interactions. *Ecology* 56: 881–892.

DeAngelis, D. L., L. Godbout, and B. J. Shuter. 1991. An individual-based approach to predicting density-dependent dynamics in smallmouth bass populations. *Ecological Modelling* 57: 91–115.

Dearing, M. D. and J. J. Schall. 1992. Testing models of optimal diet assembly by the generalist herbivorous lizard *Cnemidophorus murinus*. *Ecology* 73: 845–858.

De Roos, A. M., E. McCauley, and W. Wilson. 1991. Mobility versus density limited predator-prey dynamics on different spatial scales. *Proceedings of the Royal Society of London, Ser. B* 246: 117–122.

De Roos, A. M., J. A. J. Metz, and O. Diekmann. 1992. Studying the dynamics of structured population models: a versatile technique and its application to *Daphnia*. *American Naturalist* 139: 123–147.

Dill, L. M. 1978. An energy-based model of optimal feeding-territory size. *Theoretical Population Biology* 14: 396–429.

Dingle, H. 1994. Genetic analyses of animal migration. Pp. 145–164 in Boake, C. R. B., ed. Quantitative genetic studies of behavioral evolution. University of Chicago Press, Chicago, IL, USA.

Doebeli, M. 1996. Quantitative genetics and population dynamics. Evolution 50: 532–546.

Doucet, C. M. and J. M. Fryxell. 1993. The effect of nutritional quality on forage preference by beavers. *Oikos* 67: 201–208.

Ebenman, B. and L. Persson. eds. 1988. *Size-Structured Populations: Ecology and Evolution*. Springer Verlag, Berlin.

Ebersole, J. P. 1980. Food density and territory size: an alternative model and a test on the reef fish *Eupomacentrus leucosticus*. *American Naturalist* 115: 492–509.

Edelstein-Keshet, L. and M. D. Rausher. 1989. The effects of inducible plant defenses on herbivore populations. 1. Mobile herbivores in continuous time. *American Naturalist* 133: 787–810.

Ellner, S. and P. Turchin. 1995. Chaos in a noisy world: new methods and evidence from time-series analysis. *American Naturalist* 145: 343–375.

Elner, R. W. and R. N. Hughes. 1978. Energy maximization in the diet of the shore crab, *Carcinus maenas*. *Journal of Animal Ecology* 47: 103–116.

Endler, J. A. 1986. *Natural Selection in the Wild*. Princeton University Press, Princeton, NJ.

Engen, S. and N. C. Stenseth. 1984. A general version of optimal foraging theory: the effect of simultaneous encounters. *Theoretical Population Biology* 26: 192–204.

Fagerström, T., S. Larsson, and O. Tenow. 1987. On optimal defense in plants. *Functional Ecology* 1: 73–81.

Focardi, S., P. Marcellini, and P. Montanaro. 1996. Do ungulates exhibit a food density threshold? A field study of optimal foraging and movement patterns. *Journal of Animal Ecology* 65: 606–620.

Formanowicz, D. R., Jr. 1984. Foraging tactics of an aquatic insect: partial consumption of prey. *Animal Behaviour* 32: 774–781.

Fox, L. R. 1981. Defense and dynamics in plant-herbivore systems. *American Zoologist* 21: 853–864.

Fretwell, S. D. and H. L. Lucas. 1970. On territorial behavior and other factors influencing habitat distribution in birds. I. Theoretical development. *Acta Biotheoretica* 19: 16–36.

Fritz, R. S. and E. L. Simms. 1992. Plant resistance to herbivores and pathogens. University of Chicago Press, Chicago, IL, USA.

Fryxell, J. M. 1997. Evolutionary dynamics of habitat use. *Evolutionary Ecology* (in press).

Fryxell, J. M. and C. M. Doucet. 1993. Diet choice and the functional response of beavers. *Ecology* 74: 1297–1306.

Fryxell, J. M. and P. Lundberg. 1993. Optimal patch use and metapopulation dynamics. *Evolutionary Ecology* 7: 379–393.

Fryxell, J. M. and P. Lundberg. 1994. Diet choice and predator-prey dynamics. *Evolutionary Ecology* 8: 407–421.

Fryxell, J. M., S. M. Vamosi, R. A. Walton, and C. M. Doucet. 1994. Retention time and the functional response of beavers. *Oikos* 71: 207–214.

Gass, C. L., G. Angehr, and J. Centa. 1976. Regulation of food supply by feeding territoriality in the rofous hummingbird. *Canadian Journal of Zoology* 54: 2046–2054.

Gatto, M. 1991. Some remarks on models of plankton densities in lakes. *American Naturalist* 137: 264–267.

Getty, T. 1981. Territorial behavior of eastern chipmunks (*Tamias striatus*): encounter avoidance and spatial time-sharing. *Ecology* 62: 915–921.

Gill, G. B. 1988. Trapline foraging by hermit hummingbirds: competition for an undefended, renewable resource. *Ecology* 69: 1933–1942.

Gill, R. B. and L. Wolf. 1977. Nonrandom foraging by sunbirds in a patchy environment. *Ecology* 58: 1284–1296.

Giller, P. S. 1980. The control of handling time and its effect on the foraging strategy of a heteropteran predator, *Notonecta*. *Journal of Animal Ecology* 49: 699–712.

Gilliam, J. F. and D. F. Fraser. 1987. Habitat selection under predation hazard: a test of a model with foraging minnows. *Ecology* 68: 1856–1862.

Gilpin, M. E. 1972. Enriched predator-prey systems: theoretical stability. *Science* 177: 902–904.

Gilpin, M. E. 1975. *Group Selection on Predator-Prey Communities*. Princeton University Press, Princeton, NJ.

Gilpin, M. E. and I. Hanski. 1991. *Metapopulation Dynamics*. Academic Press, London.

Gleeson, S. K. and D. S. Wilson. 1986. Equilibrium diet: optimal foraging and prey coexistence. *Oikos* 46: 139–144.

Goodman, D. 1987. The demography of chance extinctions, Pp. 11–34 *in* M. E. Soule, ed., *Viable Populations*. Cambridge University Press, Cambridge, UK.

Goss-Custard, J. D. 1977. Optimal foraging and the size selection of worms by redshank, *Tringa totanus,* in the field. *Animal Behaviour* 25: 10–29.

Goss-Custard, J. D. and S. E. A. le V. dit Durrell. 1987*a*. Age-related effects in oystercatchers, *Haematopus ostralegus,* feeding on mussels, *Mytilus edulis*. I. Foraging efficiency and interference. *Journal of Animal Ecology* 56: 521–536.

Goss-Custard, J. D. and S. E. A. le V. dit Durrell. 1987*b*. Age-related effects in oystercatchers, *Haematopus ostralegus,* feeding on mussels, *Mytilus edulis*. III. The effect of interference on overall intake rate. *Journal of Animal Ecology* 56: 549–558.

Goss-Custard, J. D. and S. E. A. le V. dit Durrell. 1988. The effect of dominance and feeding method on the intake rates of oystercatchers, *Haematopus ostralegus,* feeding on mussels, *Mytilus edulis*. *Journal of Animal Ecology* 57: 827–844.

Goss-Custard, J. D., R. W. G. Caldow, R. T. Clarke, and A. D. West. 1995*a*. Deriving population parameters from individual variations in foraging behaviour. I. Empirical game theory distribution model of oystercatchers *Haematopus ostralegus* feeding on mussels *Mytilus edulis*. *Journal of Animal Ecology* 64: 265–276.

Goss-Custard, J. D., R. W. G. Caldow, R. T. Clarke, and A. D. West. 1995*b*. Deriving population parameters from individual variations in foraging behaviour. II. Model tests and population parameters. *Journal of Animal Ecology* 64: 277–289.

Gould, S. J. and R. C. Lewontin. 1979. The spandrels of San Marco and the Panglossian paradigm: a critique of the adaptationist programme. *Proceedings of the Royal Society of London*, Ser. B 205: 581–598.

Gray, R. D. and M. Kennedy. 1994. Perceptual constraints on optimal foraging: a reason for departures from the ideal free distribution? *Animal Behaviour* 47: 469–471.

Green, R. F. 1990. Putting ecology back into optimal foraging theory. *Comments in Theoretical Biology* 1: 387–410.

Greig-Smith, P. W. 1987. Bud-feeding by bullfinches: methods for spreading damage evenly within orchards. *Journal of Applied Ecology* 24: 49–62.

Grenfell, B. T., O. F. Price, S. D. Albon, and T. H. Clutton-Brock. 1992. Overcompensation and population cycles in an ungulate. *Nature* 355: 823–826.

Gross, J. E., L. A. Shipley, N. T. Hobbs, D. E. Spalinger and B. A. Wunder. 1993. Functional response of herbivores in food-concentrated patches—tests of a mechanistic model. *Ecology* 74: 778–791.

Gurney, W. S. C. and R. M. Nisbet. 1978. Predator–prey fluctuations in patchy environments. *Journal of Animal Ecology* 47: 85–102.

Gurney, W. S. C. and R. M. Nisbeth. 1985. Fluctuation periodicity, generation separation, and the expression of larval competition. *Theoretical Population Biology* 28: 150–180.

Haukioja, E. and T. Hakala. 1975. Herbivore cycles and periodic outbreaks: formulation of a general hypothesis. *Report of the Kevo Subarctic Research Station* 12: 1–9.

Hall, J. G. 1960. Willow and aspen in the ecology of beaver on Sagehen Creek, California. *Ecology* 41: 484–494.

Hanski, I. 1994. A practical model of metapopulation dynamics. *Journal of Animal Ecology* 63: 151–162.

Hanski, I. and M. Gilpin. 1991. Metapopulation dynamics: brief history and conceptual domain. *Biological Journal of the Linnean Society* 42: 3–16.

Hanski, I., P. Turchin, E. Korpimäki, and H. Henttonen. 1993. Population oscillations of boreal rodents: regulation by mustelid predators leads to chaos. *Nature* 364: 232–235.

Hanson, J. and L. Green. 1989. Foraging decisions: patch choice and exploitation by pigeons. *Animal Behaviour* 37: 968–986.

Harper, D. G. C. 1982. Competitive foraging in mallards: "ideal free" ducks. *Animal Behaviour* 30: 575–584.

Harper, J. L. 1977. *The Population Biology of Plants*. Academic Press, London.

Hart, D. D. 1985. Causes and consequences of territoriality in a grazing stream insect. *Ecology* 66: 404–414.

Hart, D. D. 1987. Feeding territoriality in aquatic insects: cost-benefit models and experimental tests. *American Zoologist* 27: 371–386.

Hartling, K. L. and R. C. Plowright. 1979. Foraging by bumblebees on patches of artificial flowers: a laboratory study. *Canadian Journal of Zoology* 57: 1866–1870.

Harvell, C. D. 1990. The ecology and evolution of inducible defenses. *Quarterly Review of Biology* 65: 323–340.

Hassell, M. P. 1971. Manual interference between searching insect parasites. *Journal of Animal Ecology* 40: 473–486.

Hassell, M. P. and G. C. Varley. 1969. New inductive population model for insect parasites and its bearing on biological control. *Nature* 223: 1133–1137.

Hassell, M. P., H. N. Comins, and R. M. May. 1991. Spatial structure and chaos in insect population dynamics. *Nature* 353: 255–258.

Hassell, M. P., H. N. Comins, and R. M. May. 1994. Species coexistence and self-organizing spatial dynamics. *Nature* 370: 290–292.

Hastings, A. 1977. Spatial heterogeneity and the stability of predator–prey systems. *Theoretical Population Biology* 12: 37–48.

Hastings, A. 1993. Complex interactions between dispersal and dynamics: lessons from coupled logistic equations. *Ecology* 74: 1362–1372.

Hastings, A. and K. Higgins. 1994. Persistence of transients in spatially structured ecological models. *Science* (Wash.) 263: 1133–1136.

Hastings, A. and T. Powell. 1991. Chaos in a three-species food chain. *Ecology* 72: 896–903.

Havel, J. E. and S. I. Dodson. 1984. *Chaoborus* predation on typical and spined morphs of *Daphnia pulex:* behavioral observations. *Limnology and Oceanography* 29: 487–494.

Herms, D. A. and W. J. Mattson. 1992. The dilemma of plants: to grow or defend. *Quarterly Review of Biology* 67: 283–335.

Hilborn, R. 1975. The effect of spatial heterogeneity on the persistence of predator–prey interactions. *Theoretical Population Biology* 8: 346–355.

Hixon, M. A. 1980. Food production and competitor density as the determinants of feeding territory size. *American Naturalist* 115: 510–530.

Hjältén, J., K. Danell, and L. Ericson. 1993. Effects of simulated herbivory and intraspecific competition on the compensatory ability of juvenile birches. *Ecology* 74: 1136–1142.

Hodges, C. M. 1985. Bumblebee foraging: the threshold departure rule. *Ecology* 66: 179–187.

Hodges, C. M. and L. L. Wolf. 1981. Optimal foraging bumblebees. Why is nectar left behind in flowers? *Behavioural Ecology and Sociobiology* 9: 41–44.

Hofer, H. and M. L. East. 1993*a*. The commuting system of Serengeti spotted hyaenas: how a predator copes with migratory prey. I. Social organization. *Animal Behaviour* 46: 547–557.

Hofer, H. and M. L. East. 1993*b*. The commuting system of Serengeti spotted hyaenas: how a predator copes with migratory prey. II. Intrusion pressure and commuter's space use. *Animal Behaviour* 46: 559–574.

Hogeweg, P. and B. Hesper. 1981. Two predators and one prey in a patchy environment: an application of MICMAC modelling. *Journal of Theoretical Biology* 93: 411–432.

Holling, C. S. 1959. The components of predation as revealed by a study of small-mammal predation of the European pine sawfly. *Canadian Entomologist* 91: 293–320.

Holmes, E. E. 1993. Are diffusion models too simple: a comparison with telegraph models of invasion. *American Naturalist* 142: 779–795.

Holmgren, N. 1995. The ideal free distribution of unequal competitors: predictions from a behaviour-based functional response. *Journal of Animal Ecology* 64: 197–212.

Holt, R. D. 1983. Optimal foraging and the form of the predator isocline. *American Naturalist* 122: 521–541.

Holt, R. D. 1985. Population dynamics in two-patch environments: some anomalous consequences of an optimal habitat distribution. *Theoretical Population Biology* 28: 181–208.

Hoskinson, R. L. and L. D. Mech. 1976. White-tailed deer migration and its role in wolf predation. *Journal of Wildlife Management* 40: 429–441.

Houston, A. I., R. H. McCleery, and N. B. Davies. 1985. Territory size, prey renewal and feeding rates: interpretation of observations on the pied wagtail (*Motacilla alba*) by simulation. *Journal of Animal Ecology* 54: 227–239.

Houston, A. I., J. N. McNamara, and J. M. C. Hutchinson. 1993. General results concerning the trade-off between gaining energy and avoiding predation. *Philosophical Transactions of the Royal Society of London,* Ser. B, 341: 375–397.

Howe, H. F. and R. B. Primack. 1975. Differential seed dispersal by birds of the tree *Casearia nitidia* (Flacourtiaceae). *Biotropica* 7: 278–283.

Hubbard, S. F., R. M. Cook, J. G. Glover, and J. J. D. Greenwood. 1982. Apostatic selection as an optimal foraging strategy. *Journal of Animal Ecology* 51: 625–633.

Huffaker, C. B. 1958. Experimental studies in predation: dispersion factors and predator–prey oscillations. *Hilgardia* 27: 343–383.

Hughes, R. N. 1979. Optimal diets under the energy maximization premise: the effects of recognition time and learning. *American Naturalist* 113: 209–221.

Hughes, R. N. 1990. *Behavioural Models of Food Selection.* Springer-Verlag, Berlin, Germany.

Hughes, R. N. and M. T. Burrows. 1991. Diet selection by dogwhelks in the field: an example of constrained optimization. *Animal Behaviour* 42: 47–55.

Hughes, R. N. and M. I. Croy. 1993. An experimental analysis of frequency-dependent predation (switching) in the 15-spined stickleback, *Spinachia spinachia. Journal of Animal Ecology* 62: 341–352.

Hughes, T. P. 1984. Population dynamics based on individual size rather than age: a general model with a reef coral example. *American Naturalist* 123: 778–795.

Huntly, N. J., A. T. Smith, and B. L. Ivins. 1986. Foraging behavior of the pika (*Ochotona princeps*) with comparisons of grazing versus haying. *Journal of Mammalogy* 67: 139–148.

Hsu, S. B., S. P. Hubbell, and P. Waltman. 1978. A contribution to the theory of competing predators. *Ecological Monographs* 48: 337–349.

Illius, A. W., S. D. Albon, J. M. Pemberton, I. J. Gordon, and T. H. Clutton-Brock. 1995. Selection for foraging efficiency during a population crash in Soay sheep. *Journal of Animal Ecology* 64: 481–492.

Ives, A. R. and A. P. Dobson. 1987. Antipredator behavior and the population dynamics of simple predator–prey systems. *American Naturalist* 130: 431–447.

Janetos, A. C. and B. J. Cole. 1981. Imperfectly optimal animals. *Behavioral Ecology and Sociobiology* 9: 203–209.

Janzen, D. H. 1970. Herbivores and the number of tree species in tropical forests. *American Naturalist* 104: 501–528.

Janzen, D. H. 1971. Seed predation by animals. *Annual Review of Ecology and Systematics,* 2: 465–492.

Jenkins, S. H. 1980. A size-distance relation in food selection by beavers. *Ecology* 61: 740–746.

Johnson, D. F. and G. Collier. 1989. Patch choice and meal size of foraging rats as a function of the profitability of food. *Animal Behaviour* 38: 285–297.

Juliano, S. A. and F. M. Williams. 1985. On the evolution of handling time. *Evolution* 39: 212–215.

Karban, R. and J. H. Myers. 1989. Induced plant responses to herbivory. *Annual Review of Ecology and Systematics* 20: 331–348.

Kareiva, P. 1987. Habitat fragmentation and the stability of predator–prey interactions. *Nature* 326: 388–390.

Kareiva, P. 1990. Population dynamics in spatially-complex environments: theory and data. *Philosophical Transactions of the Royal Society of London,* Ser. B 330: 175–190.

Kareiva, P. and G. M. Odell. 1987. Swarms of predators exhibit "preytaxis" if individual predators use area-restricted search. *American Naturalist* 130: 233–270.

Kareiva, P. and U. Wennergren. 1995. Connecting landscape patterns to ecosystem and population processes. *Nature* 373: 299–302.

Kaspari, M. and A. Joern. 1993. Prey choice by three insectivorous grassland birds: reevaluating opportunism. *Oikos* 68: 414–430.

Kidd, N. A. C. and A. D. Mayer. 1983. The effect of escape responses on the stability of insect host–parasite models. *Journal of Theoretical Biology* 104: 275–287.

Kingsland, S. E. 1985. Modeling nature. University of Chicago Press, Chicago, IL, USA.

Kitting, C. L. 1980. Herbivore–plant interactions of individual limpets maintaining a mixed diet of intertidal marine algae. *Ecological Monographs* 50: 527–550.

Koch, A. L. 1974. Competitive coexistence of two predators utilizing the same prey under constant environmental conditions. *Journal of Theoretical Biology* 44: 387–395.

Kodric-Brown, A. and J. H. Brown. 1978. Influence of economics, interspecific competition, and sexual dimorphism on territoriality of migrant rufous hummingbirds. *Ecology* 59: 285–296.

Korona, R. 1989. Ideal free distribution of unequal competitors can be determined by the form of competition. *Journal of Theoretical Biology* 138: 347–352.

Korona, R. 1990. Travel costs and ideal free distribution of ovipositing female flour beetles, *Tribolium confusum. Animal Behaviour* 40: 186–187.

Korpimäki, E., K. Norrdahl, and J. Valkama. 1994. Reproductive investment under fluctuating predation risk: microtine rodents and small mustelids. *Evolutionary Ecology* 8: 357–368.

Kotler, B. P. 1992. Behavioral resource depression and decaying perceived risk of predation in two species of coexisting gerbils. *Behavioral Ecology and Sociobiology* 30: 239–244.

Krebs, J. R. and N. B. Davies. 1981. *An Introduction to Behavioural Ecology.* Blackwell Scientific Publications, Oxford, UK.

Krebs, J. R. and R. H. McCleery. 1984. Optimization in behavioural ecology. Pp. 91–121 *in* J. R. Krebs and N. B. Davies, eds., *Behavioural Ecology.* Blackwell, Oxford, UK.

Krebs, J. R., J. T. Erichsen, and M. I. Webber. 1977. Optimal prey selection in the great tit (Parus major). *Animal Behavior* 25: 30–38.

Krivan, V. 1996. Optimal foraging and predator–prey dynamics. *Theoretical Population Biology* 49: 265–290.

Krivan, V. 1997. Dynamic ideal free distribution: effects of optimal patch choice on predator-prey dynamics. *American Naturalist* 149: 164–178.

Lacher, T. E., M. R. Willig, and M. A. Mares. 1982. Food preference as a function of resource abundance with multiple prey types: an experimental analysis of optimal foraging theory. *American Naturalist* 120: 297–316.

Lande, R. 1993. Risk of extinction from demographic and environmental stochasticity and random catastrophes. *American Naturalist* 142: 911–927.

Latto, J. 1994. Evidence for a self-thinning rule in animals. *Oikos* 69: 531–534.

Lawlor, L. R. and J. Maynard Smith. 1976. The coevolution and stability of competing species. *American Naturalist* 110: 79–99.

Lawton, J. H., J. R. Beddington, and R. Bonser. 1974. Switching in invertebrate predators. Pp. 141–158 *in* M. B. Usher and M. H. Williamson, eds., *Ecological Stability*. Chapman & Hall, London.

Lechowicz, M. J. and G. Bell. 1992. The ecology and genetics of fitness in forest plants II. Microscale heterogeneity of the edaphic environment. *Journal of Ecology* 79: 687–696.

Lefkovitch, L. P. 1965. The study of population growth in organisms grouped by stages. *Biometrics* 21: 1–18.

Leigh, E. G. 1981. The average lifetime of a population in a varying environment. *Journal of Theoretical Biology* 90: 213–239.

Leon, J. A. and D. Tumpson. 1975. Competition between two species for two complementary or substitutable resources. *Journal of Theoretical Biology* 50: 185–201.

Levin, S. A. 1974. Dispersion and population interactions. *American Naturalist* 108: 207–228.

Levins, R. 1979. Coexistence in a variable environment. *American Naturalist* 114: 765–783.

Levins, R. and D. Culver. 1971. Regional coexistence of species and competition between rare species. *Proceedings of the National Academy of Sciences U.S.A.* 68: 1246–1248.

Lewis, M. A. and J. D. Murray. 1993. Modelling territoriality and wolf–deer interactions. *Nature* 366: 738–740.

Lima, S. L. 1985. Maximizing feeding efficiency and minimizing time exposed to predators: a trade-off in the black-capped chickadee. *Oecologia* (Berl.) 66: 60–67.

Lima, S. L. and L. M. Dill. 1990. Behavioral decisions made under the risk of predation: a review and prospectus. *Canadian Journal of Zoology* 68: 619–640.

Lima, S. L., T. J. Valone, and T. Caraco. 1985. Foraging-efficiency-predation-risk trade-off in the grey squirrel. *Animal Behaviour* 33: 155–165.

Lonsdale, W. M. 1990. The self-thinning rule: dead or alive? *Ecology* 71: 1373–1388.

Lotka, A. J. 1925. *Elements of Physical Biology*. Williams and Wilkins, Baltimore, MD.

Lotka, A. J. 1932. Contribution to the mathematical theory of capture. I. Conditions for capture. *Proceedings of the National Academy of Science, U.S.A.* 18: 172–178.

Lucas, J. R. 1983. The role of foraging time constraints and variable prey encounter in optimal diet choice. *American Naturalist* 122: 191–209.

Lucas, J. R. 1985. Partial prey consumption by antlion larvae. *Animal Behaviour* 33: 945–958.

Lucas, J. R. and P. M. Waser. 1989. Defense through exploitation: a Skinner box for tropical rain forests. *Trends in Ecology and Evolution* 4: 62–63.

Lundberg, P. 1988. Functional response of a small mammalian herbivore: the disc equation revisited. *Journal of Animal Ecology* 57: 999–1006.

Lundberg, P. and M. Åström. 1990*a*. Low nutritive quality as a defense against optimally foraging herbivores. *American Naturalist* 135: 547–562.

Lundberg, P. and M. Åström. 1990*b*. Functional response of optimally foraging herbivores. *Journal of Theoretical Biology* 144: 367–377.

Lundberg, P. and K. Danell. 1990. Functional response of browsers: tree exploitation by moose. *Oikos* 58: 378–384.

Lundberg, P. and J. M. Fryxell. 1995. Expected population density vs. productivity in ratio-dependent and prey-dependent models. *American Naturalist* 146: 153–161.

Lundberg, P. and R. T. Palo. 1994. Resource use, plant defense, and optimal digestion in ruminants. *Oikos* 68: 224–228.

Lundberg, S., J. Järemo, and P. Nilsson. 1994. Herbivory, inducible defence and population oscillations: a preliminary theoretical analysis. *Oikos* 71: 537–540.

MacArthur, R. H. and E. R. Pianka. 1966. On optimal use of a patchy environment. *American Naturalist* 100: 603–609.

Malcolm, S. B. 1992. Prey defence and predator foraging, Pp. 458–475 *in* M. J. Crawley, ed., *Natural Enemies*. Blackwells, Oxford, UK.

Mangel, M. 1992. Rate maximizing and state variable theories of diet selection. *Bulletin of Mathematical Biology* 54: 413–422.

Mangel, M. and C. Clark. 1986. Towards a unified foraging theory. *Ecology* 67: 1127–1138.

Mangel, M. and C. W. Clark. 1988. *Dynamic Modelling in Behavioral Ecology*. Princeton University Press, Princeton, NJ.

Mangel, M. and B. D. Roitberg. 1992. Behavioral stabilization of host-parasite population dynamics. *Theoretical Population Biology* 42: 308–320.

Manly, B., L. McDonald, and D. Thomas. 1993. Resource selection by animals. Chapman & Hall, London.

Manseau, M. 1996. Relationships between the rivière George caribou herd and its summer range. Unpublished Ph.D. dissertation, Université Laval, Québec.

Mares, M. A. and T. E. Lacher. 1987. Social spacing in small mammals: patterns of individual variation. *American Zoologist* 27: 293–306.

Martinsen, G. D., J. H. Cushman, and T. G. Whitham. 1990. Impact of pocket gopher disturbance on plant diversity in a shortgrass prairie community. *Oecologia* (Berl.) 83: 132–138.

Matsuda, H. and T. Namba. 1989. Co-evolutionarily stable community structure in a patchy environment. *Journal of Theoretical Biology* 136: 229–243.

Matsuda, H., P. A. Abrams, and M. Hori. 1993. The effect of adaptive anti-predator behavior on exploitative competition and mutualism between predators. *Oikos* 68: 549–559.

May, R. M. 1972. Limit cycles in predator-prey communities. *Science* 177: 900–902.

May, R. M. 1973. *Stability and Complexity in Model Ecosystems*. Princeton University Press, Princeton, NJ.

Maynard Smith, J. 1982. *Evolution and the Theory of Games*. Cambridge University Press, Cambridge, UK.

Maynard Smith, J. and G. P. Price. 1973. The logic of animal conflicts. *Nature* 246: 15–18.

McArthur, C., C. T. Robbins, A. E. Hagerman, and T. A. Hanley. 1993. Diet selection by a ruminant generalist browser in relation to plant chemistry. *Canadian Journal of Zoology* 71: 2236–2243.

McCann, K., and P. Yodzis. 1994. Biological conditions for chaos in a three-species food chain. *Ecology* 75: 561–564.

McCauley, E., W. W. Murdoch, and S. Watson. 1988. Simple models and variation in plankton densities among lakes. *American Naturalist* 132: 383–403.

McCauley, E., W. G. Wilson, and A. M. de Roos. 1993. Dynamics of age-structured and spatially structured predator-prey interactions: individual-based models and population-level formulations. *American Naturalist* 142: 412–442.

McFarland, D. J. and A. I. Houston. 1981. Quantitative ethology: the state space approach. Pitman, London, UK.

McGinley, M. A. and T. G. Whitham. 1985. Central place foraging by beavers (*Castor canadensis*): a test of foraging predictions and the impact of selective feeding on the growth form of cottonwoods (*Populus fremontii*). *Oecologia* (Berl.) 66: 558–562.

McLaren, B. E. and R. O. Peterson. 1994. Wolves, moose, and tree rings on Isle Royale. *Science* (Wash.) 266: 1555–1558.

McLaughlin, J. F. and J. Roughgarden. 1991. Pattern and stability in predator–prey communities: how diffusion in spatially variable environments affects the Lotka–Volterra model. *Theoretical Population Biology* 40: 148–172.

McNair, J. N. 1987. The effect of variability on the optimal size of a feeding territory. *American Zoologist* 27: 249–258.

McNamara, J. M. and A. I. Houston. 1982. Short term behavior and lifetime fitness. Pp. 60–87 *in* D. J. McFarland, ed., Functional ontogeny. Pitman, London, UK.

McNamara, J. M. and A. I. Houston. 1986. The common currency for behavioral decisions. *American Naturalist* 127: 358–378.

McNamara, J. M. and A. I. Houston. 1987. Partial preferences and foraging. *Animal Behaviour* 35: 1084–1099.

McNamara, J. M. and A. I. Houston. 1990. State-dependent ideal free distributions. *Evolutionary Ecology* 4: 293–311.

Mech, L. D. 1994. Buffer zones of territories of gray wolves as regions of intraspecific strife. *Journal of Mammalogy* 75: 199–202.

Metz, J. A. J. and O. Diekmann (eds). 1986. *The Dynamics of Physiologically Structured Populations.* Springer-Verlag, Berlin.

Metz, J. A. J., A. M. de Roos, and F. van den Bosch. 1988. Population models incorporating physiological structure: A quick survey of the basic concepts and an application to size-structured population dynamics in waterfleas, Pp. 106–126 *in* B. Ebenman and L. Persson, eds., *Size-Structured Populations.* Springer-Verlag, Berlin.

Milinski, M. 1979. An evolutionarily stable feeding strategy in sticklebacks. *Zeitschrift fur Tierpsychologie* 51: 36–40.

Milinski, M. and R. Heller. 1978. Influence of a predator on the optimal foraging behaviour of sticklebacks (*Gasterosteus aculeatus* L.). *Nature* 275: 642–644.

Milton, K. 1979. Factors influencing leaf choice by howler monkeys: a test of some hypotheses of food selection by generalist herbivores. *American Naturalist* 114: 362–378.

Mimura, M. and J. D. Murray. 1978. On a diffusive prey-predator model which exhibits patchiness. *Journal of Theoretical Biology* 75: 249–262.

Mitchell, W. A. and J. S. Brown. 1990. Density-dependent harvest rates by optimal foragers. *Oikos* 57: 180–190.

Moody, A. L. and A. I. Houston. 1995. Interference and the ideal free distribution. *Animal Behaviour* 49: 1065–1072.

Morista, M. 1952. Habitat preference and evaluation of environment of an animal: experimental studies on the population density of an ant lion, *Glenuroides japonicus* M'L. *Physiological Ecology* 5: 1–16.

Morris, D. W. 1987. Tests of density-dependent habitat selection in a patchy environment. *Ecological Monographs* 57: 269–281.

Morris, D. W. 1988. Habitat-dependent population regulation and community structure. *Evolutionary Ecology* 2: 253–269.

Morris, D. W. 1989. Density-dependent habitat selection: testing the theory with fitness data. *Evolutionary Ecology* 3: 80–94.

Murdoch, W. W. 1973. The functional response of predators. *Journal of Applied Ecology* 14: 335–341.

Murdoch, W. W. 1977. Stabilizing effects of spatial heterogeneity in predator-prey systems. *Theoretical Population Biology* 11: 252–273.

Murdoch, W. W. 1994. Population regulation in theory and practice. *Ecology* 75: 271–287.

Murdoch, W. W. and A. Oaten. 1975. Predation and population stability. *Advances in Ecological Research* 9: 2–131.

Murdoch, W. W., S. Avery, and M. E. B. Smyth. 1975. Switching in predatory fish. *Ecology* 56: 1094–1105.

Myers, J. P., P. G. Connors, and F. A. Pitelka. 1979*a*. Territory size in wintering sanderlings: the effects of prey abundance and intruder density. *The Auk* 96: 551–561.

Myers, J. P., P. G. Connors, and F. A. Pitelka. 1979*b*. Optimal territory size and the sanderling: compromises in a variable environment, Pp. 135–158 *in* A. C. Kamil and T. D. Sergent, eds. *Mechanisms of Foraging Behavior.* Garland Press, NY.

Nachman, G. 1987. Systems analysis of acarine predator-prey interactions. II. The role of spatial processes in system stability. *Journal of Animal Ecology* 56: 267–281.

Nachman, G. 1991. An acarine predator-prey metapopulation system inhabiting greenhouse cucumbers. *Biological Journal of the Linnean Society* 42: 285–303.

Nevonen, S. and E. Haukioja. 1991. The effect of inducible resistance in host foliage on birch-feeding herbivores. Pp. 277–292 *in* D. W. Tallamy and M. J. Raupp, eds., *Phytochemical Induction by Herbivores.* Academic Press, NY.

Newman, J. A., A. J. Parsons, and A. Harvey. 1992. Not all sheep prefer clover: diet selection revisited. *Journal of Agricultural Science* 119: 275–283.

Nisbet, R. M. and W. S. C. Gurney. 1982. *Modelling Fluctuating Populations.* Wiley, Chichester, UK.

Nisbet, R. M., C. J. Briggs, W. S. C. Gurney, W. W. Murdoch, and A. Stewart-Oaten. 1993. Two-patch metapopulation dynamics. Pp. 125–135 *in* S. A. Levin, T. M. Powell, and J. H. Steele, eds., *Patch Dynamics.* Springer-Verlag, Berlin.

Nonacs, P. 1993. Is satisficing an alternative to optimal foraging theory? *Oikos* 67: 371–375.

Nonacs, P. and L. M. Dill. 1990. Mortality risk vs. food quality trade-offs in a common currency: ant patch preferences. *Ecology* 71: 1886–1892.

Norberg, Å. 1988. Self-thinning of plant populations dictated by packing density and individual growth geometry and relationships between animal population density and body mass governed by metabolic rate. Pp. 259–279 *in* B. Ebenman and L. Persson, eds., *Size-Structured Populations.* Springer-Verlag, Berlin.

Noy-Meir, I. 1975. Stability of grazing systems: an application of predator–prey graphs. *Journal of Ecology* 63: 459–481.

Noy-Meir, I. 1976. Rotational grazing in a continuously growing pasture: a simple model. *Agricultural Systems* 1: 87–112.

Oksanen, L. and P. Lundberg. 1995. Optimization of reproductive effort and foraging time in mammals: the influence of resource level and predation risk. *Evolutionary Ecology* 9: 45–56.

Okubo, A. 1980. *Diffusion and Ecological Problems: Mathematical Models. Biomathematics* 10, Springer-Verlag, Berlin.

Orians, G. H. and N. E. Pearson. 1979. On the theory of central place foraging, Pp.

155–177 *in* D. J. Horn, G. R. Stairs, and R. D. Mitchell, eds., *Analysis of Ecological Systems.* Ohio State University Press, Columbus, Ohio.

Osenberg, C. W. and G. G. Mittelbach. 1989. Effects of body size on the predator-prey interaction between pumpkinseed sunfish and gastropods. *Ecological Monographs* 59: 405–432.

Owen-Smith, N. 1993. Evaluating optimal diet models for an African browsing ruminant: the kudu: how constraining are the assumed constraints? *Evolutionary Ecology* 7: 499–524.

Owen-Smith, N. 1994. Foraging responses of kudus to seasonal changes in resources: elasticity in constraints. *Ecology* 75: 1050–1062.

Owen-Smith, N. and P. Novellie. 1982. What should a clever ungulate eat? *American Naturalist* 119: 151–178.

Pacala, S. W. and J. A. Silander, Jr. 1985. Neighborhood models of plant population dynamics. I. Single-species models of annuals. *American Naturalist* 125: 385–411.

Pacala, S. W. and J. A. Silander, Jr. 1990. Field tests of neighborhood population dynamics models of two annual weed species. *Ecological Monographs* 60: 113–134.

Parker, G. A. and W. A. Sutherland. 1986. Ideal free distributions when individuals differ in competitive ability: phenotype-limited ideal free models. *Animal Behavior* 34: 1222–1242.

Pastorok, R. A. 1981. Prey vulnerability and size selection by *Chaoborus* larvae. *Ecology* 62: 1311–1324.

Paszkowski, C. A., W. M. Tonn, J. Pirionen, and I. J. Holopainen. 1989. An experimental study of body size and food size selection in crucian carp, *Carassius carrasius. Environmental Biology of Fishes* 24: 275–286.

Paton, D. C. and F. L. Carpenter. 1984. Peripheral foraging by territorial rufous hummingbirds: defense by exploitation. *Ecology* 65: 1808–1819.

Penning, S. C., M. T. Nadeau, and V. J. Paul. 1993. Selectivity and growth of the generalist herbivore *Dolabella auricularia* feeding upon complementary resources. *Ecology* 74: 879–890.

Persson, L. 1985. Optimal foraging: the difficulty of exploiting different feeding strategies simultaneously. *Oecologia* (Berl.), 67: 338–341.

Persson, L. 1987. The effects of resource availability and distribution on size class interactions in perch (*Perca fluviatilis*). *Oikos* 48: 148–160.

Persson, L. 1988. Asymmetries in predatory and competitive interactions in fish populations. Pp. 203–218 *in* L. Persson and B. Ebendman, eds., *Size-Structured Populations: Ecology and Evolution.* Springer-Verlag, Berlin.

Persson, L. 1991. Behavioural response to predators reverses the outcome of competition between prey species. *Behavioural Ecology and Sociobiology* 28: 101–105.

Persson, L. and S. Diehl. 1990. Mechanistic individual-based approaches in the population/community ecology of fish. *Annales Zoologici Fennici* 27: 165–182.

Persson, L. and L. A. Greenberg. 1990. Optimal foraging and habitat shift in perch (*Perca fluviatilis*) in a resource gradient. *Ecology* 71: 1699–1713.

Persson, L., J. Bengtsson, B. A. Menge, and M. A. Power. 1996. Productivity and consumer regulation—concepts, patterns, and mechanisms. Pp. 396–434 in G. A. Polis and K. D. Winemiller, eds., Food Webs, Chapman & Hall, NY, USA.

Peterman, R. M., W. C. Clark, and C. S. Holling. 1979. The dynamics of resilience: shifting stability domains in fish and insect systems. Pp. 321–341 *in* R. M. Anderson,

B. D. Turner, and L. R. Taylor, eds., *Population Dynamics* (20th Symposium of the British Ecological Society). Blackwell, Oxford.

Pickup, G. 1994. Modelling patterns of defoliation by grazing animals in rangelands. *Journal of Applied Ecology* 31: 231–246.

Pickup, G. and Chewings, V. H. 1988. Estimating the distribution of grazing and patterns of cattle movement in a large arid zone paddock: an approach using animal distribution models and Landsat imagery. *International Journal of Remote Sensing* 9: 1469–1490.

Pierce, G. J. and J. G. Ollason. 1987. Eight reasons why optimal foraging theory is a complete waste of time. *Oikos* 49: 111–118.

Pimm, S. L. 1982. Food webs. Chapman & Hall, London, UK.

Pimm, S. L., M. L. Rosenzweig, and W. Mitchell. 1985. Competition and food selection: field tests of a theory. *Ecology* 66: 798–807.

Pimm, S. L., H. L. Jones, and J. Diamond. 1988. On the risk of extinction. *American Naturalist* 132: 757–785.

Pinkowski, B. 1983. Foraging behavior of beavers (*Castor canadensis*) in North Dakota. *Journal of Mammalogy* 64: 312–314.

Polis, G. 1984. Age structure component of niche width and intraspecific resource partitioning: Can age groups function as ecological species? *American Naturalist* 123: 541–564.

Possingham, H. P. 1989. The distribution and abundance of resources encountered by a forager. *American Naturalist* 133: 42–60.

Pulliam, H. R. 1974. On the theory of optimal diets. *American Naturalist* 108: 59–74.

Pulliam, H. R. 1975. Diet optimization with nutrient constraints. *American Naturalist* 109: 765–768.

Pyke, G. H. 1980. Optimal foraging in hummingbirds: testing the marginal value theorem. *American Zoologist* 18: 739–752.

Ranta, E. and V. Nuutinen. 1985. Foraging by the smooth newt (*Trituris vulgaris*) on zooplankton: functional responses and diet choice. *Journal of Animal Ecology* 54: 275–293.

Rapport, D. J. 1980. Optimal foraging for complementary resources. *American Naturalist* 116: 324–346.

Raubenheimer, D. and S. J. Simpson. 1993. The geometry of compensatory feeding in the locust. *Animal Behaviour* 45: 953–964.

Real, L. A. 1994. Ecological genetics. Princeton University Press, Princeton, NJ, USA.

Recer, G. M., W. U. Blanckenhorn, J. A. Newman, E. M. Tuttle, M. L. Withiam, and T. Caraco. 1987. Temporal resource variability and the habitat-matching rule. *Evolutionary Ecology* 1: 363–378.

Rechten, C., M. Avery, and A. Stevens. 1983. Optimal prey selection: why do Great Tits show partial preferences? *Animal Behaviour* 31: 576–584.

Rice, J. A., T. J. Miller, K. A. Rose, L. B. Crowder, E. A. Marshall, A. S. Trebitz, and D. L. DeAngelis. 1993. Growth rate variation and larval survival: inferences from an individually based size-dependent predation model. *Canadian Journal of Fisheries and Aquatic Sciences* 50: 133–142.

Riessen, H. P. 1992. Cost-benefit model for the induction of antipredator defense. *American Naturalist* 140: 349–362.

Ritchie, M. E. 1988. Individual variation in the ability of Columbian ground squirrels to select an optimal diet. *Evolutionary Ecology* 2: 232–252.

Roff, D. 1992. *The Evolution of Life-Histories.* Chapman & Hall, New York.

Rosenzweig, M. L. 1971. Paradox of enrichment: destabilization of exploitation ecosystems in ecological time. *Science* (Wash.) 171: 385–387.

Rosenzweig, M. L. 1981. A theory of habitat selection. *Ecology* 62: 327–335.

Rosenzweig, M. L. 1986. Hummingbird isolegs in an experimental system. *Behavioral Ecology and Sociobiology* 19: 313–322.

Rosenzweig, M. L. 1991. Habitat selection and population interactions: the search for mechanism. *American Naturalist* 137: S5–28.

Rosenzweig, M. L. and R. H. MacArthur. 1963. Graphical representation and stability conditions of predator-prey interactions. *American Naturalist* 97: 209–223.

Ruxton, G. D., W. S. C. Gurney, and A. M. de Roos. 1992. Interference and generation cycles. *Theoretical Population Biology* 42: 235–253.

Sabelis, M. W., O. Diekmann, and V. A. A. Jansen. 1991. Metapopulation persistence despite local extinction: predator–prey patch models of the Lotka–Volterra type. *Biological Journal of the Linnean Society* 42: 267–283.

Saloniemi, I. 1993. A coevolutionary predator–prey model with quantitative characters. *American Naturalist* 141: 880–896.

Schaeffer, R. L. and W. Mendenhall. 1975. *Introduction to Probability: Theory and Applications.* Duxbury Press, North Scituate, MA.

Scheel, D. 1993. Profitability, encounter rates, and prey choice of African lions. *Behavioural Ecology* 4: 90–97.

Schluter, D. 1981. Does the theory of optimal diets apply in complex environments? *American Naturalist* 118: 139–147.

Schmitz, O. J., D. S. Hik, and A. R. E. Sinclair. 1992. Plant chemical defense and twig selection by snowshoe hares: an optimal foraging perspective. *Oikos* 65: 295–300.

Schoener, T. W. 1971. Theory of feeding strategies. *Annual Review of Ecology and Systematics* 2: 369–404.

Schoener, T. W. 1979. Generality of the size-distance relation in models of optimal foraging. *American Naturalist* 114: 902–914.

Schoener, T. W. 1983. Simple models of optimal feeding-territory size: a reconciliation. *American Naturalist* 121: 608–629.

Schoener, T. 1986. Mechanistic approaches to community ecology: a new reductionism. *American Zoologist* 26: 81–106.

Schoener, T. W. 1987. Time budgets and territory size: some simultaneous optimization models for energy maximizers. *American Zoologist* 27: 259–291.

Schwinning, S. and M. L. Rosenzweig. 1990. Periodic oscillations in an ideal-free predator-prey distribution. *Oikos* 59: 85–91.

Sih, A. 1980. Optimal behavior: can foragers balance two conflicting demands? *Science* (Wash.) 210: 1041–1043.

Sih, A. 1982. Foraging strategies and the avoidance of predation by an aquatic insect, *Notonecta hoffmanni. Ecology* 63: 786–796.

Sih, A. 1984. Optimal behavior and density-dependent behavior. *American Naturalist* 123: 314–326.

Sih, A. 1987. Prey refuges and predator–prey stability. *Theoretical Population Biology* 31: 1–12.

Simms, E. L. 1992. Cost of plant resistance to herbivory. Pp. 392–425 *in* R. S. Fritz and E. L. Simms, eds., *Plant Resistance to Herbivores and Pathogens—Ecology, Evolution and Genetics.* University of Chicago Press, Chicago.

Sinclair, A. R. E. 1985. Does interspecific competition or predation shape the African ungulate community? *Journal of Animal Ecology* 54: 899–918.

Sinclair, A. R. E. 1989. Population regulation in animals, Pp. 197–241 *in* J. M. Cherrett, ed., *Ecological Concepts*. British Ecological Society Symposium 26, Blackwell, Oxford, UK.

Sinclair, A. R. E., J. M. Gosline, G. Holdsworth, C. J. Krebs, S. Boutin, J. N. M. Smith, R. Boonstra, and M. Dale. 1993. Can the solar cycle and climate synchronize the snowshoe hare cycle in Canada? Evidence from tree rings and ice cores. *American Naturalist* 141: 173–198.

Skellam, J. G. 1951. Random dispersal in theoretical populations. *Biometrika* 38: 196–218.

Skogsmyr, I. and T. Fagerström. 1992. The cost of anti-herbivory defence: an evaluation of some ecological and physiological factors. *Oikos* 64: 451–457.

Skulason, S. and T. B. Smith. 1995. Resource polymorphisms in vertebrates. *Trends in Ecology and Evolution* 10: 366–370.

Skulason, S., S. S. Snorrason, D. Otas, and D. L. G. Noakes. 1993. Genetically based differences in foraging behaviour among sympatric morphs of arctic char (Pisces: Salmonidae). *Animal Behaviour* 45: 1179–1192.

Smuts, G. L. 1978. Interrelations between predators, prey, and their environment. *BioScience* 28: 316–320.

Stearns, S. 1992. *The Evolution of Life Histories.* Oxford University Press, Oxford.

Stearns, S. and J. Koella. 1986. The evolution of phenotypic plasticity in life history traits: predictions for norms of reaction for age- and size-at-maturity. *Evolution* 40: 893–913.

Stearns, S. C. and P. Schmid-Hempel. 1987. Evolutionary insights should not be wasted. *Oikos* 49: 118–125.

Stein, R. A. 1977. Selective predation, optimal foraging, and the predator–prey interaction between fish and crayfish. Ecology 58: 1237–1253.

Stein, R. A. and J. J. Magnuson. 1976. Behavioral response of crayfish to a fish predator. *Ecology* 57: 751–761.

Stemberger, R. S. and J. J. Gilbert. 1987. Multi-species induction of morphological defenses in the rotifer *Keratella testudo. Ecology* 68: 370–378.

Stenseth, N. C. 1981. Optimal food selection: some further considerations with special reference to the grazer-hunter distinction. *American Naturalist* 117: 457–475.

Stenseth, N. C. and L. Hansson. 1979. Optimal food selection: a graphic model. *American Naturalist* 113: 373–389.

Stephens, D. W. 1985. How important are partial preferences? *Animal Behaviour* 33: 667–669.

Stephens, D. W. and S. R. Dunbar. 1993. Dimensional analysis in behavioral ecology. *Behavioral Ecology* 4: 172–183.

Stephens, D. and J. R. Krebs. 1986. *Foraging Theory.* Princeton University Press, Princeton, NJ.

Stewart, W. J. 1994. *Introduction to the Numerical Solution of Markov Chains.* Princeton University Press, Princeton, NJ.

Streams, F. A. 1994. Effects of prey size on attack components of the functional response by *Notonecta undulata. Oecologia* (Berl.) 98: 57–63.

Sutherland, W. J. and G. A. Parker. 1985. Distribution of unequal competitors. Pp. 255–273 *in* R. M. Sibly and R. H. Smith, eds., *Behavioral Ecology: Ecological Consequences of Adaptive Behaviour.* Blackwell, Oxford.

Sutherland, W. A. and G. A. Parker. 1992. The relationship between continuous input and interference models of ideal free distributions with unequal competitors. *Animal Behaviour* 44: 345–355.

Sutherland, W. J., C. R. Townsend, and J. M. Patmore. 1988. A test of the ideal free distribution with unequal competitors. *Behavioral Ecology and Sociobiology* 23: 51–53.

Taitt, D. E. 1988. The dynamics of stand development: a general stand model applied to Douglas fir. *Canadian Journal of Forest Research* 18: 696–702.

Tanner, J. T. 1975. The stability and the intrinsic growth rates of prey and predator populations. *Ecology* 56: 855–867.

Tansky, M. 1978. Switching effect in predator–prey system. *Journal of Theoretical Biology* 70: 263–271.

Taylor, A. D. 1991. Studying metapopulation effects in predator–prey systems. *Biological Journal of the Linnean Society* 42: 305–323.

Taylor, R. J. 1984. *Predation.* Chapman & Hall, New York.

Taylor, R. J. 1988. Territory size and location in animals with refuges: influence of predation risk. *Evolutionary Ecology* 2: 95–101.

Taylor, R. J. and P. J. Pekins. 1991. Territory boundary avoidance as a stabilizing factor in wolf-deer interactions. *Theoretical Population Biology* 39: 115–128.

Tilman, G. D. 1982. *Resource Competition and Community Structure.* Princeton University Press, Princeton, NJ.

Tilman, D., R. M. May, C. L. Lehman, and M. A. Nowak. 1994. Habitat destruction and the extinction debt. *Nature* 371: 65–66.

Turchin, P. 1991. Translating foraging movements in heterogeneous environments into the spatial distribution of foragers. *Ecology* 72: 1253–1266.

Turchin, P. and P. Kareiva. 1989. Aggregation in *Aphis varians:* an effective strategy for reducing predation risk. *Ecology* 70: 1008–1016.

Turchin, P. and A. D. Taylor. 1992. Complex dynamics in ecological time series. *Ecology* 73: 289–305.

Turelli, M., J. H. Gillespie, and T. W. Schoener. 1982. The fallacy of the fallacy of the averages in ecological optimization theory. *American Naturalist* 119: 879–884.

Tyler, J. A. and K. A. Rose. 1994. Individual variability and spatial heterogeneity in fish population models. *Review of Fish Biology and Fisheries* 4: 91–123.

Vail, S. G. 1993. Scale-dependent responses to resource spatial pattern in simple models of consumer movement. *American Naturalist* 141: 199–216.

van Baalen, M. 1994. Evolutionary stability and the persistence of predator-prey systems. Ph.D. Thesis, University of Amsterdam.

van Baalen, M. and M. W. Sabelis. 1993. Coevolution of patch selection strategies of predator and prey, and the consequences for ecological stability. *American Naturalist* 142: 646–670.

Van Orsdol, K. G. 1984. Foraging behaviour and hunting success of lions in Queen Elizabeth National Park, Uganda. *African Journal of Ecology* 22: 79–99.

Verlinden, C. and R. H. Wiley. 1989. The constraints of digestive rate: an alternative model of diet selection. *Evolutionary Ecology* 3: 264–273.

Vermeij, G. J. 1982. Unsuccessful predation and evolution. *American Naturalist* 120: 701–720.

Volterra, V. 1928. Variations and fluctuations of the number of individuals in animal species living together. *Journal du Conseil international pour l'Exploration de la Mer* 3: 3–51.

Waddington, K. D. 1982. Optimal diet theory: sequential and simultaneous encounter models. *Oikos* 39: 278–280.

Waldbauer, G. P. 1988. Asynchrony between Batesian mimics and their models. *American Naturalist* 131: S103–S121.

Waldbauer, G. P., J. G. Sternburg, and C. T. Maier. 1977. Phenological relationships of wasps, bumblebees, their mimics, and insectivorous birds in an Illinois sand area. *Ecology* 58: 583–591.

Walls, M. and M. Ketola. 1989. Effects of predator-induced spines on individual fitness in *Daphnia pulex*. *Limnology and Oceanography* 34: 390–396.

Waltman, P. 1983. *Competition Models in Population Biology*. Society for Industrial and Applied Mathematics, Philadelphia, PA.

Walton, W. E., N. G. Hairston, Jr., and J. K. Wetterer. 1992. Growth-related constraints on diet selection by sunfish. *Ecology* 73: 429–437.

Ward, D. 1992. The role of satisficing in foraging theory. Oikos 63: 312–317.

Ward, D. 1993. Foraging theory, like all other fields of science, needs multiple working hypotheses. *Oikos* 67: 376–378.

Ward, D. and S. Saltz. 1994. Foraging at different spatial scales: Dorcas gazelles foraging for lilies in the Negev Desert. *Ecology* 75: 48–58.

Watkinson, A. R. 1980. Density-dependence in single-species populations of plants. *Journal of Theoretical Biology* 83: 345–357.

Werner, E. E. 1986. Amphibian metamorphosis: growth rates, predation risk, and the optimal size at transformation. *American Naturalist* 128: 319–341.

Werner, E. E. 1992. Individual behavior and higher-order species interactions. *American Naturalist* 140: S5–S32.

Werner, E. E. and B. R. Anholt. 1993. Ecological consequences of the trade-off between growth and mortality rates mediated by foraging activity. *American Naturalist* 142: 242–272.

Werner, E. E. and J. F. Gilliam. 1984. The ontogenetic niche and species interactions in size-structured populations. *Annual Review of Ecology and Systematics* 5: 393–425.

Werner, E. E. and D. J. Hall. 1974. Optimal foraging and the size selection of prey by the bluegill sunfish (*Lepomis macrochirus*). *Ecology* 55: 1042–1052.

Werner, E. E. and D. J. Hall. 1988. Ontogenetic habitat shifts in bluegill: the foraging rate-predation risk trade-off. *Ecology* 69: 1352–1366.

Werner, E. E. and G. G. Mittlebach. 1981. Optimal foraging: field tests of diet choice and habitat switching. *American Zoologist* 21: 813–829.

Werner, E. E., G. G. Mittlebach, D. J. Hall, and J. F. Gilliam. 1983*a*. Experimental tests of optimal habitat use in fish: the role of relative habitat profitability. *Ecology* 64: 1525–1539.

Werner, E. E., J. F. Gilliam, D. J. Hall, and G. G. Mittlebach. 1983*b*. An experimental test of the effects of predation risk on habitat use in fish. *Ecology* 64: 1540–1548.

Werner, P. A. 1975. Predictions of fate from rosette size in teasel (*Dipsacus fullonum* L.). *Oecologia* (Berl.) 20: 197–201.

Werner, P. A. and H. Caswell. 1977. Population growth rates and age vs. stage distribution models for teasel (*Dipsacus sylvestris* Huds.). *Ecology* 58: 1103–1111.

West, L. 1988. Prey selection by the tropical snail *Thais melones:* a study of interindividual variation. *Ecology* 69: 1839–1854.

Westoby, M. 1974. An analysis of diet selection by large generalist herbivores. *American Naturalist* 108: 290–304.

Westoby, M. 1978. What are the biological bases of varied diets? *American Naturalist* 112: 627–631.

Wetterer, J. K. and C. J. Bishop. 1985. Planktivore prey selection: the reactive field volume model vs. the apparent size model. *Ecology* 66: 457–464.

White, J. 1977. Generalization of self-thinning of plant populations. *Nature* 268: 373.

White, J. and J. L. Harper. 1970. Correlated changes in plant size and number in plant populations. *Journal of Ecology* 58: 467–485.

Wilson, D. S. and J. Yoshimura. 1994. On the coexistence of specialists and generalists. *American Naturalist* 144: 692–707.

Witteman, G. J., A. Redfearn, and S. L. Pimm. 1990. The extent of complex population changes in nature. *Evolutionary Ecology* 4: 173–183.

Yoccoz, N. G., S. Engen, and N. C. Stenseth. 1993. Optimal foraging: the importance of environmental stochasticity and accuracy in parameter estimation. *American Naturalist* 141: 139–157.

Yodzis, P. 1989. *Introduction to Theoretical Ecology.* Harper and Row, New York.

Yodzis, P. and S. Innes. 1992. Body size and consumer-resource dynamics. *American Naturalist* 139: 1151–1175.

Ziegler, B. P. 1977. Persistence and patchiness of predator–prey systems induced by discrete event population exchange mechanisms. *Journal of Theoretical Biology* 67: 687–713.

Index